11th Anglo-French Physical Acoustics Conference

AA008588

(AFPAC 2012)

Journal of Physics: Conference Series Volume 457

Brighton, United Kingdom
18-20 January 2012

ISBN: 978-1-62993-095-4
ISSN: 1742-6588

Printed from e-media with permission by:

Curran Associates, Inc.
57 Morehouse Lane
Red Hook, NY 12571

Some format issues inherent in the e-media version may also appear in this print version.

Copyright© by the Institute of Physics
All rights reserved.

Printed by Curran Associates, Inc. (2013)

For permission requests, please contact the Institute of Physics
at the address below.

Institute of Physics
Dirac House, Temple Back
Bristol BS1 6BE UK

Phone: 44 1 17 929 7481
Fax: 44 1 17 920 0979

techtracking@iop.org

Additional copies of this publication are available from:

Curran Associates, Inc.
57 Morehouse Lane
Red Hook, NY 12571 USA
Phone: 845-758-0400
Fax: 845-758-2634
Email: curran@proceedings.com
Web: www.proceedings.com

TABLE OF CONTENTS

Preface

The 11th Anglo-French Physical Acoustics Conference (AFPAC) was held in Brighton, UK on 18–20 January 2012. This event, which is an annual collaboration between the Physical Acoustics Group (PAG) of the Institute of Physics and the Groupe d'Acoustique Physique, Sous-marine et UltraSonore (GAPSUS) of the Société Française d'Acoustique, successfully achieved its main aim of being a small, friendly meeting of high scientific quality, welcoming younger researchers and PhD students and covering a broad range of subjects in Acoustics.

The participants heard 44 excellent presentations covering an exciting and diverse range of subjects, from audio acoustics to guided waves in composites and from phononic crystals to ultrasound surgery. As is the custom at these meetings, four prominent invited speakers set the pace for the event; these were Keith Attenborough (The Open University, UK), Claire Prada (Institut Langevin, France), David Moore (University of Nottingham, UK) and Philippe Roux (IS Terre, France).

The submission of manuscripts for publication in the proceedings was, as in previous years, on a voluntary basis and in these proceedings we present 11 peer reviewed papers. Due to some unforeseen problems there has been a longer than planned delay in preparing these proceedings, for which the Editors sincerely apologise to the authors and the community.

Nader Saffari, Mike Lowe and Alain Lhémery

Content from this work may be used under the terms of the Creative Commons Attribution 3.0 licence. Any further distribution of this work must maintain attribution to the author(s) and the title of the work, journal citation and DOI.
Published under licence by IOP Publishing Ltd

Peer review statement

All papers published in this volume of *Journal of Physics: Conference Series* have been peer reviewed through processes administered by the proceedings Editors. Reviews were conducted by expert referees to the professional and scientific standards expected of a proceedings journal published by IOP Publishing.

Content from this work may be used under the terms of the Creative Commons Attribution 3.0 licence. Any further distribution of this work must maintain attribution to the author(s) and the title of the work, journal citation and DOI.
Published under licence by IOP Publishing Ltd

Journal of Physics: Conference Series **457** (2013) 012001 doi:10.1088/1742-6596/457/1/012001

Theoretical characterisation of the radial and translational motion of coated microbubbles under acoustic excitation

C J Harfield[1], N C Ovenden[2] , G Memoli[3] , P H Jones[4] and E P J Stride[1]

[1] Institute of Biomedical Engineering, Department of Engineering Science, University of Oxford, OX3 7DQ, UK
[2]Department of Mathematics, University College London, Gower Street, London, WC1E 6BT, UK
[3] National Physical Laboratory, Teddington, Middlesex, TW11 0LW, UK
[4] Department of Physics and Astronomy, University College London, Gower Street, London, WC1E 6BT, UK

E-mail: caroline.harfield@eng.ox.ac.uk

Abstract. Ultrasound contrast agents, in the form of coated microbubbles, are a powerful tool in current diagnostic imaging. Given their sensitive dynamic response they also have the potential to be used for quantitative measurements of the properties of the surrounding tissue (*e.g.* percentage perfusion or blood pressure). For this potential to be realised, however, the theoretical descriptions of bubble behaviour, in particular the constitutive equations for the microbubble shell, need to be improved and a method needs to be developed for the accurate characterisation of individual bubbles. In this paper the first steps are taken towards deriving a complete model for the coupled radial and translational motion of a coated bubble. It is then shown that with this model the bubble can be characterised by a unique set of parameters describing the bubble shell corresponding to its viscous and elastic response. This uniqueness will enable the model to be used to interpret experimental data and quantify these parameters for which accurate values are currently lacking but which are critical to predicting bubble response and hence enabling advanced diagnostic applications.

1. Introduction

Microbubbles are currently the most effective type of ultrasound contrast agent available. They consist of a gas core stabilised by a surfactant or polymer shell (figure.1.1) and have been in clinical use for more than two decades [1]. They scatter ultrasound much more effectively than red blood cells on account of the large contrast in density and compressibility between them and the surrounding blood. Most importantly, even at moderate ultrasound pressures, microbubbles undergo nonlinear volumetric oscillations in response to ultrasound excitation. Consequently, the scattered, or more correctly re-radiated signal, contains components other than those in the incident ultrasound beam. At typical

[1] To whom any correspondence should be addressed.

Content from this work may be used under the terms of the Creative Commons Attribution 3.0 licence. Any further distribution of this work must maintain attribution to the author(s) and the title of the work, journal citation and DOI.
Published under licence by IOP Publishing Ltd

diagnostic pressures these are harmonics and/or fractional harmonics of the incident frequency. This allows the blood vessels carrying microbubbles to be differentiated from the surrounding tissue, which makes contrast enhanced ultrasound a very useful modality for imaging tissue perfusion [2].

For a given ultrasound field, the signal emitted will be different for different sizes of bubble, different coating materials and for different environments, for example changes in local viscosity, temperature or pressure. This means that microbubbles have the potential to be used for quantitative interrogation of the surrounding tissue (e.g. percentage perfusion or blood pressure), in addition to image contrast enhancement. To achieve this, however, accurate characterisation of the bubble is needed, in particular to determine the normally unknown shell properties.

This paper is concerned with the development of theoretical models for bubble characterisation under acoustic excitation, and proving that the bubble can be described by a unique set of shell properties.

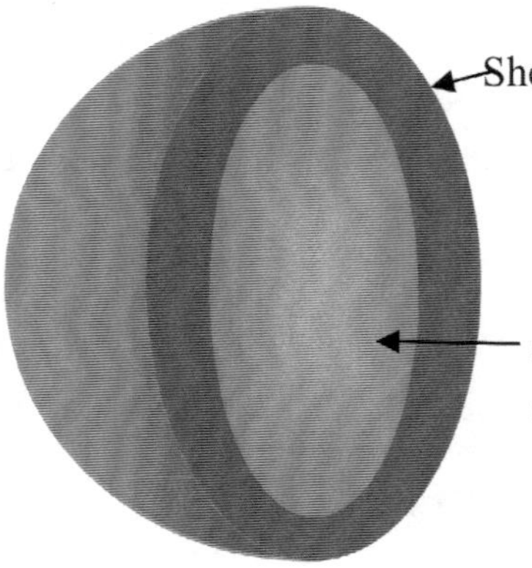

Figure 1.1 Diagram showing the different layers of a microbubble, the gas core has a radius typically between 0.5-8 μm and the shell thickness is between 1-100 nm.

2. Existing Models

2.1. Radial model

Under ultrasound excitation, the spatially and temporally varying pressure in the surrounding medium causes microbubbles to oscillate volumetrically on account of their compressibility. Bubbles may also undergo translational motion due to the steady pressure gradient produced by the propagating wave. In a standing wave a bubble will travel to the nearest node or antinode depending on its size and the acoustic frequency.

Many equations neglect the translational motion and instead only describe the radial dynamics. These equations are usually derivatives of the Rayleigh-Plesset or, more correctly, the Rayleigh-Plesset-Noltingk-Neppiras-Poritsky (RPNNP) equation[3]:

$$R\ddot{R} + \frac{3}{2}\dot{R}^2 = \frac{1}{\rho_L}\left(\left(p_0 + \frac{2\sigma}{R} - p_V\right)\left(\frac{R_0}{R}\right)^{3\gamma} + p_v - \frac{2\sigma}{R} - \frac{4\mu_L\dot{R}}{R} - p_0 + p_{ac}\right) \quad 2.1$$

In equation (2.1) R is the bubble radius with initial stationary value R_0, $\dot{R}$ is the velocity of the bubble wall, $\ddot{R}$ is the acceleration of the bubble wall, p_v is vapour pressure, γ is the polytropic constant, σ is the surface tension, μ_L is the viscosity of the liquid, p_{ac} is the pressure due to the acoustic wave, ρ_L is the density of the liquid and p_0 is the hydrostatic pressure of the liquid.

However the RPNNP equation only describes the motion of an uncoated bubble. Church [4] was the first author to publish a model for an ultrasound contrast agent with a finite thickness viscoelastic shell on the microbubble surface. Church's model does not include any thermal or radiative dissipation effects, but the paper did indicate how they could be incorporated.

Hoff *et al*. [5] took the limit of Church's model for an infinitely thin shell to obtain the governing equation

$$R\ddot{R}+\frac{3}{2}\dot{R}^2=\frac{1}{\rho_L}\left(-p_0+p_{ac}+p_G(R)-\frac{4\dot{R}}{R}\mu_L-\frac{12\mu_S d_S R_0^2\dot{R}}{R^4}-\frac{12G_s d_s R_0^2}{R^3}\left(1-\frac{R_0}{R}\right)\right)\quad(2.2)$$

that is, in fact, mathematically equivalent to the equation derived by de Jong *et al.*[6-8]. In this equation G_s represents the shear modulus, d_s, the thickness of the shell, μ_s the viscosity of the shell, and $p_G(R)=p_0\left(\frac{R_0}{R}\right)^{3\gamma}$ is the pressure due to the gas inside the bubble.

Glazman [9] and Morgan *et al.* [10] built on previous studies of oceanic bubbles, to develop models that consider the vibrations of surfactant coated microbubbles, which treated the shell as an absorbed surface layer instead of a discrete shell. Although these treatments are essentially the same as [6], they use different functional relationships to describe the effective elasticity and viscosity of the shell. However there was an error in Glazman's derivation in [9]which was propagated by Morgan but later corrected by Marmottant *et al.* [11] who produced a model that described the effect of the shell through an area dependent interfacial tension with a constant surface viscosity.

Stride [12] developed a model, equation (2.3), in which interfacial tension and viscosity are modelled as continuous functions of surface molecular concentration, Γ, obeying the experimentally derived functions of a power law and exponential relationship respectively. These variable terms replace the more conventional elastic terms seen in previous models and yield the following equation for an encapsulated bubble.

$$R\ddot{R}+\frac{3}{2}\dot{R}^2=\frac{1}{\rho_L}\left(-p_0+p_{ac}+p_G(R)-\frac{4\dot{R}}{R}\mu_L-\frac{4\dot{R}}{R^2}\eta_{so}e^{\frac{zR_y^2}{(R^2-R_y^2)}}\right.$$
$$\left.-\frac{2}{R}\left(\sigma_0+\frac{K\Gamma_0^{y+1}}{(y+1)}\left(1-\left(\frac{R_0}{R}\right)^{2(y+1)}\right)\right)\right)\quad(2.3)$$

Where Γ_0 is the initial molecular concentration, σ_0 is surface tension at equilibrium, K is the Power law constant, y is the power law exponent, z the surface viscosity exponent and η_{so} is the surface viscosity constant.

2.2 Coupled Translational and Radial Models

Interest in bubble translational motion initially started with the investigation by King [13] into the effect of acoustic radiation pressure on a sphere in a standing wave. Watanabe and Kukita [14] solved the radial and translational equations simultaneously for an uncoated bubble of mass m_b and volume $V(t)$, using the RPNNP for the radial displacement $R(t)$ alongside the following equation for the bubble's translation from the origin $x(t)$: using the RPNNP for the radial and equation (2.4) for the translational motion.

$$\ddot{x}=\frac{1}{m_b}\left(-V\frac{dp_{ac}}{dx}-\frac{1}{2}\rho_L\frac{d}{dt}(V\,u_r)-\frac{1}{2}\rho_L|u_r|u_r\,A\,C_d\right)\quad(2.4)$$

Where first term on the right hand side is the primary Bjerknes force, the second is the added mass and the third is the drag term. It is noteworthy that the governing equation for translational displacement is dependent on the relative velocity of the liquid $u_r=u_b-u_L$, where $u_b=\dot{x}$ and u_L is the liquid velocity, leading to a drag term proportional to $A=\pi R^2$ (the so-called projected cross-sectional area of the bubble) and the following drag coefficient

$$C_d = 27Re^{-0.78} \quad (2.5)$$

where *Re* is the Reynolds number defined by

$$Re = \frac{2\,\rho_L\,u_r R}{\mu} \qquad (2.6)$$

Doinikov [15] solved the radial and translational motion problem initially for an uncoated bubble using a Lagrangian approach to derive coupled equations.

$$R\ddot{R} + \frac{3}{2}\dot{R}^2 = \frac{1}{\rho_L}\left(\left(p_0 + \frac{2\sigma}{R_0}\right)\left(\frac{R_0}{R}\right)^{3\gamma} - \frac{2\sigma}{R} - p_0 + p_{ac} - \frac{4\mu_L \dot{R}}{R}\right) + \frac{\dot{x}^2}{4} \qquad (2.7)$$

$$\ddot{x} + \frac{3\dot{R}\dot{x}}{R} = \frac{3F_{ex}}{2\pi\rho_L R^3} \qquad (2.8)$$

Where F_{ex} represents the external forces (including the viscous drag and primary Bjerknes forces).

This derivation ignores the mass of the bubble and the relative velocity is treated inconsistently, being included in the viscous drag term but not in the kinetic energy of the liquid. A new term, $\frac{\dot{x}^2}{4}$ was however included in the radial equation which coupled equations (2.7) and (2.8) together. This term is also referred to as the slip velocity [16] and had been noted previously [15], but ignored because it is often negligibly small; however this is not always the case and should therefore be included for accuracy.

Later Doinikov and Dayton [17], re-derived the coupled equations using the same method as above, but for a shelled bubble using the Church equation (2.2) to describe the radial dynamics including the coupling term along with

$$\ddot{x} = \frac{1}{\left(m_b + \frac{2}{3}\pi R_2^3 \rho_L\right)}\left(-\frac{4\pi R_2^3}{3}\frac{d\,p_{ac}(x,t)}{dx} - 6\pi\mu_L R\dot{x} - 2\pi\rho_L\left(R_2^2 \dot{R}\dot{x}\right)\right) \quad (2.9)$$

to describe the translational motion, where R_2 is the outside radius of the bubble including the shell. We note here that equation (2.9) can be shown to be equivalent to that of equation (2.4) if the relative motion between the liquid and the bubble is ignored.

Finally Dayton *et al.* [18] subsequently published a model ,which also included some inconsistent terms in particular accounting for the added mass due to the oscillations of the bubble twice.

3. Derivation of Model

In reviewing the previous equations a number of inconsistencies were discovered. As above, these included the frame of reference used and the treatment of the relative velocity between the bubble and the liquid. In [15], the relative velocity is included in the drag force but not in the added mass or slip velocity in the radial term; whilst the model in [17] neglects the relative velocity of the bubble with respect to the liquid completely. Therefore a new version is derived in this section, following a discussion of the underlying assumptions.

3.1 Assumptions

The assumptions made in the derivation of the model are as follows:

1. Spherical symmetry is maintained during microbubble oscillation.
2. External forces such as gravity and buoyancy can be ignored as they are very small in comparison to other forces experienced by the bubble.
3. There are no changes in the bubble properties during oscillation, i.e. equilibrium radius, shell and liquid properties remain constant (unless otherwise stated). This includes the assumption

that the bubble remains intact and there is negligible gas diffusion over the period of oscillation [19].

4. Thermal damping and radiation are taken to be negligible on account of the small amplitudes of oscillation being considered.
5. Due to the gas having very low density and viscosity compared with that of the surrounding liquid, the inertial and viscous contributions of the gas can be taken to be negligible.
6. The acoustic wavelength of the ultrasound is much larger than the radius of the bubble. Hence pressure variations across the bubble can be regarded as negligible.
7. Vapour inside the bubble is negligible; i.e. it is filled purely with gas. This is because the majority of microbubbles are manufactured under gas saturated conditions [20] and it is assumed that the shell would inhibit inwards diffusion after bubble formation.
8. The thermal diffusivity of the filling gas must remain constant during oscillations for the above assumptions to be consistent. Therefore it is also assumed to exhibit ideal and polytropic behaviour.
9. The surrounding fluid is taken to be infinite, initially static, homogeneous, incompressible and Newtonian.

3.2 Derivation

Following the approach adopted by Doinikov [15, 17] and starting from the Lagrangian equation

$$\frac{d}{dt}\frac{dL}{d\dot{q}_i} - \frac{dL}{dq_i} = -\frac{dF}{d\dot{q}_i} \quad (3.1)$$

Where L is the Lagrangian function defined by $L{=}T{-}U$, T is the kinetic energy, U is the potential energy and q_i and $\dot{q}_i$ are the generalised coordinates and velocities respectively. The F is the dissipative function which will take account of the dissipative effects of the liquid viscosity.
The total kinetic energy function in the frame of reference of the bubble centre is given by

$$T = 2\pi\rho_L R^3\left(\dot{R}^2 + \frac{{u_r}^2}{6}\right) \quad (3.2)$$

and hence includes the relative velocity of the liquid, $u_r = u_b - u_L$ where the bubble velocity $u_b = \dot{x}$ and the liquid velocity u_L, can be described by

$$u_L = \frac{p_a}{\rho_L c_s}\cos(\omega\, t)\cos(k\, x) \quad (3.3)$$

with the angular frequency $\omega = 2\pi f$, the wavenumber $k = \frac{2\,\pi\, f}{c_s}$, where p_a is the pressure amplitude, c_s is the speed of sound and f is the frequency of the acoustic wave. It should be noted that equation(3.2) does not include the kinetic energy of the bubble as in [17]. This is because in the frame of reference used this term does not exist.
The total potential energy is given by the sum of the potential energy of the gas within the bubble, the potential energy of the surface tension and the potential energy of the external pressure, leading to the following expression given the initial volume of the bubble V_0,

$$U = \frac{p_0 V_0}{\gamma - 1}\left(\frac{R_0}{R}\right)^{3(\gamma-1)} + 4\pi R^2\sigma + \frac{4\pi}{3}R^3[p_0 + p_{ac}] \quad (3.4)$$

Equation (3.4) does not contain the viscous forces and the dissipative function of the liquid is of the form

$$F = \int_{V_L} f_L dV \qquad (3.5)$$

where V_L is the volume occupied by the liquid, and f_L is the density of the dissipative function. Defining ξ_L as the bulk viscosity coefficient and v_{ij} as the rate-of-strain tensor of the liquid, the integrand of equation (3.5) can be written

$$f_L = \mu_L \left(v_{ij} - \frac{1}{2}\delta_{ij} v_{kk}\right)^2 + \frac{1}{2}\xi_L v_{kk}^2 \qquad (3.6)$$

Moreover, since the liquid is incompressible, $v_{kk} = \mathbf{\Delta v} = 0$, equation (3.6) becomes

$$f_L = \mu_L \left(v_{ij}\right)^2 \qquad (3.7)$$

and by integrating $\left(v_{ij}\right)^2$ over V_L , the volume of the liquid, it is found that

$$F_L = 8\pi\mu_L R\dot{R}^2 + 3\pi\mu_L R {u_r}^2 \qquad (3.8)$$

Substituting equations (3.2),(3.4) and(3.8) into equation (3.1) for generalised coordinate q =R and x respectively then yields

$$R\ddot{R} + \frac{3}{2}\dot{R}^2 = \frac{1}{\rho_L}\left(\left(p_0 + \frac{2\sigma}{R_0}\right)\left(\frac{R_0}{R}\right)^{3\gamma} - \frac{2\sigma}{R} - p_0 - p_{ac} - \frac{4\mu_L \dot{R}}{R}\right) + \frac{u_r^2}{4} \qquad (3.9)$$

$$\ddot{x} = \frac{1}{\left(\frac{2}{3}\pi\rho_L R^3\right)}\left(-2\pi\rho_L \dot{R} R^2 u_r + \frac{2}{3}\pi\rho_L R^3 \dot{u}_L - \frac{4}{3}\pi R^3 \frac{dp_{ac}(x,t)}{dx} - 6\pi\mu_L R u_r + \frac{\pi\, \rho_l R^3}{3}\frac{du_r^2}{dx}\right) \qquad (3.10)$$

As can be seen this gives the RPNNP equation with the slip velocity included and an equation for the translational motion that is the same as that derived in[15] apart from the relative motion being used to describe the velocity of the bubble in the flow as well as the inclusion of the final $\frac{du_r^2}{dx}$ term. [17] would also produce the same equation if the relative velocity were treated correctly and the frame of reference consistently applied.

In most applications, the $\frac{du_r^2}{dx}$ term in equation(3.10) is negligibly small and makes no difference to the bubble's behaviour. For equation (3.10) it should also be noted that the derivation assumes Stokes' law as the drag term, which only holds for Reynolds number Re << 1.

The model needs to include a shell for the purposes of characterising a coated bubble. It was shown by Doinikov and Dayton [17] that the translational term and coupling term are unaltered by the presence of a shell as described by Church's model[4]. As Hoff's model [5] represents the thin shell limit of Church's equation, the translational equation above can still be used, yielding the following equation for the radial dynamics

$$R\ddot{R} + \frac{3}{2}\dot{R}^2 = \frac{1}{\rho_L}\left(p_0 + p_{ac} + p_G(R) - \frac{4\dot{R}}{R}\mu_L - \frac{12\mu_S d_S R_0^2 \dot{R}}{R^4} - \frac{12 G_s d_s R_0^2}{R^3}\left(1 - \frac{R_0}{R}\right)\right) + \frac{u_r^2}{4} \qquad (3.11)$$

Equation (3.11) was verified against the analytical solution of the Hoff model [5]. In figure 3.1 it is shown that difference between using the slip velocity and neglecting it with an example of when this term starts to become important. This graph was produced in MatLabRb2011a using ode45 with the values in Table 1 and run for 50 acoustic cycles. The shell parameters were based on those of SonoVue found in [21] .

Table1. Showing the parameters used to produce the figure.3.1.

Parameter	Value
R_0	1μm
p_0	100kPa
μ_s	2.5Pa.s
μ_L	$8.9\text{x}10^{-4}$Pa.s
Σ	$7.2\text{x}10^{-2}$N.m^{-1}
ρ_L	998 kg .m^{-3}
G_s	$122\text{x}10^{6}$ Pa
d_s	1.5nm
c_s	1500m.s^{-1}
ρ_g	1.83kg m^{-3}
Γ	1.4

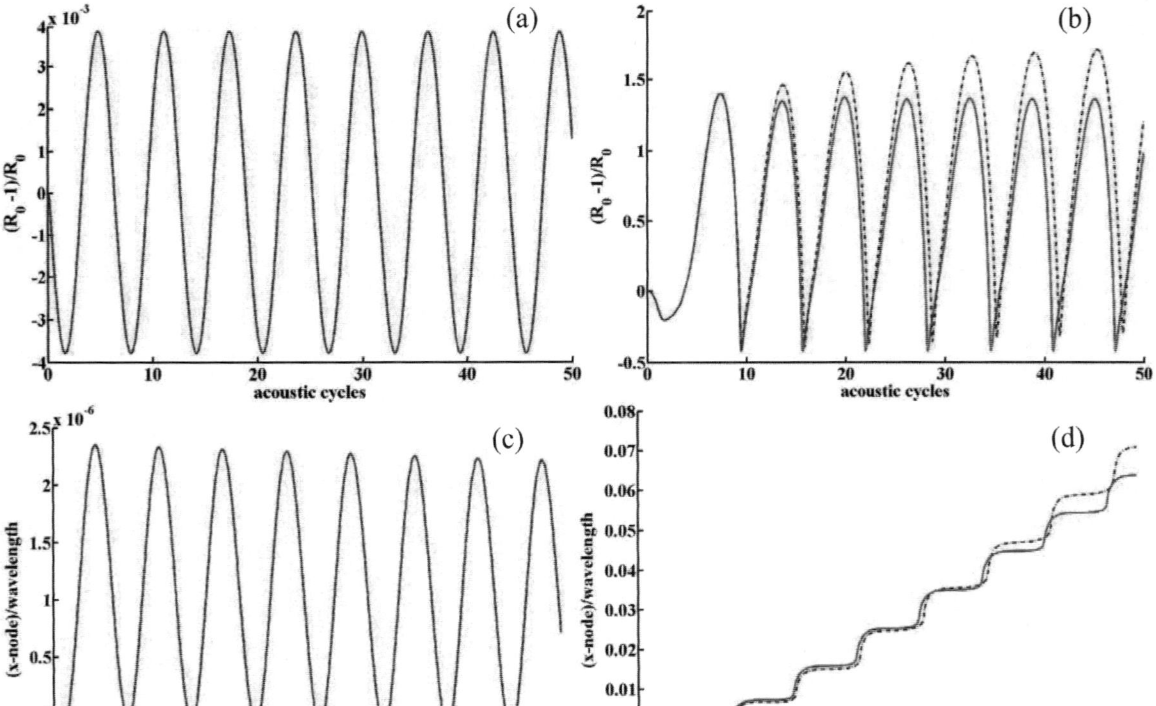

Figure 3.1 Comparison between using the set of equations (3.10) and (3.11) represented by the blue dotted line, against equation (2.2) and (3.10) represented by the red line which neglects the coupling term in the radial equation. Where (a) is the radial, (c) is the translational motion of the bubble with p_{ac}=10kPa and frequency 0.7MHz , (b) is the radial and (d) the translational motion of the bubble with p_{ac}=1.1MPa and frequency 4.9MHz.

4. Uniqueness of the Equation

In order for characterisation of the shell to be useful it is essential that a unique set of shell parameters can be determined through fitting experimental data to a theoretical model. Assume for instance that the variation in radius *R(t)* of an encapsulated bubble driven by a known external ultrasound pressure field $p_{ac}(t)$ can be accurately measured during an experiment. If everything about the bubble and the surrounding medium is known aside from the properties of the shell then one can assume the behaviour of *R(t)* to be governed by a model such as that given in (2.2) and then attempt to find values of the shell's shear modulus G_S, the shell thickness, d_s, and the shell viscosity, μ_s, that fit the experimental *R(t)* profile accurately. However, the question remains over whether such parameter values are unique for the bubble that was measured experimentally.

To answer the uniqueness question, equation (2.2) can be written in the following way

$$\rho_L\left(R\ddot{R}+\frac{3}{2}\dot{R}^2\right)+(-p_G(R)+p_0-p)+\frac{4\dot{R}}{R}\mu_L-\frac{\rho_L u_r^2}{4} = -\frac{12\mu_S d_S R_S^2\dot{R}}{R^4}-\frac{12G_s d_S R_S^2}{R^3}\left(1-\frac{R_0}{R}\right) \quad (4.1)$$

where only terms containing the unknown shell parameters are retained on the right-hand side. The three parameters μ_s, d_s, G_s can then be reduced to just two unknowns, $G_s d_s = v$ *and* $\mu_s d_s = u$ as they only appear as these two products, noting that d_s can be measured independently. For this proof it shall be assumed that these parameter values are constant over an oscillation and their uniqueness will be proved by contradiction. Hence, we initially assume that values v and u are not unique and thus there must exist at least two sets of values v_1, u_1 and v_2, u_2 that, when substituted into (4.1), identically replicate the experimental measurement of *R(t)*. Redefining the left hand side of equation (4.1) to be equal to a function $F(\ddot{R},\dot{R},R)$, it must therefore be the case that .

$$F(\ddot{R},\dot{R},R) = -\frac{12\,R_0^2\dot{R}}{R^5}u_1-\frac{12\,R_0^2}{R^4}\left(1-\frac{R_0}{R}\right)v_1 \quad (4.2a)$$

$$F(\ddot{R},\dot{R},R) = -\frac{12\,R_0^2\dot{R}}{R^5}u_2-\frac{12\,R_0^2}{R^4}\left(1-\frac{R_0}{R}\right)v_2 \quad (4.2b)$$

Eliminating $F(\ddot{R},\dot{R},R)$ from these equations and rearranging then yields the ordinary first-order differential equation for *R(t)*,

$$\dot{R}+\frac{R(v_1-v_2)}{(u_1-u_2)}=\frac{R_0(v_1-v_2)}{(u_1-u_2)} \quad (4.3)$$

Letting $\frac{(v_1-v_2)}{(u_1-u_2)}=\beta$ and solving subsequently reveals the general solution to be

$$R(t)=R_0+\frac{C}{\beta}e^{-\beta t} \quad (4.4)$$

where *C* is an unknown constant, which can be set by the initial condition. Equation (4.4) can be interpreted in the following way: if the varying bubble radius *R(t)* in the experiment does not exhibit exponential decay to the value R_0 then it must be the case that $u_1 = u_2$ and $v_1 = v_2$. As a consequence, any data-fitted parameter values must be unique (so long as d_s can be measured separately).

Following similar logic it can also be proven that if values of parameters η_{so} and $K\Gamma_0$ are fitted to experimental data of $R(t)$ in equation (2.3) with power-law exponent y=0 and surface viscosity exponent z=0, then those fitted values are unique. In this case the governing equation is given by :

$$\rho_L\left(R\ddot{R}+\frac{3}{2}\dot{R}^2\right)+p_0-p_{ac}-p_G(R)+\frac{4\dot{R}}{R}\mu_L=-\frac{4\dot{R}}{R^3}\eta_{so}-\mathrm{K}\Gamma_0\frac{3}{\mathrm{R}^2}\left(1-\frac{\mathrm{R}_0^2}{\mathrm{R}^2}\right) \quad (4.5)$$

Hence, as before by taking $\alpha_1=\eta_{so}$, $\beta_1=K\Gamma_0$ and $\alpha_2=\eta_{so}$, $\beta_2=K\Gamma_0$ as two sets of parameter values that accurately fit the experimental data, these two sets can be substituted into (4.5) and, on eliminating the common left-hand side of the equation in each case, yields

$$R\frac{dR}{dt}+\Omega R^2=\Omega \mathrm{R}_0^2$$

for $\Omega=\frac{3(\beta_1-\beta_2)}{4(\alpha_1-\alpha_2)}$. The general solution to this differential equation is then $R(t)=\sqrt{(R_0^2+2Ce^{-2\Omega t})}$ and hence unless $R^2(t)$ exponentially decays in this manner to R_0^2 in the experimental data then the fitted parameter values must be unique.

5. Conclusions

Previous descriptions of the coupled radial and translational motion of a bubble have contained inconsistent treatments of the relative velocity of the bubble and been inconsistent with the frames of reference used. A derivation to include these has been presented here and these effects have been shown to be non-negligible, particularly at high pressures.

It has been proven that a unique set of parameters describing a microbubble coating can be derived by fitting experimental data to the theoretical model. This will enable it to be used effectively for bubble characterisation.

References

[1] Cosgrove, D., Ultrasound contrast agents: An overview. *European Journal of Radiology*, 2006. **60**(3): p. 324-330.

[2] Stride, E.P. and C.C. Coussios, Cavitation and contrast: the use of bubbles in ultrasound imaging and therapy. *Proceedings of the Institution of Mechanical Engineers Part H-Journal of Engineering in Medicine*, 2010. **224**(H2): p. 171-191.

[3] Lauterborn, W., Numerical Investigation of Nonlinear Oscillations of Gas-Bubbles in Liquids. *Journal of the Acoustical Society of America*, 1976. **59**(2): p. 283-293.

[4] Church, C.C., The Effects of An Elastic Solid-Surface Layer on the Radial Pulsations of Gas-Bubbles. *Journal of the Acoustical Society of America*, 1995. **97**(3): p. 1510-1521.

[5] Hoff, L., P.C. Sontum, and J.M. Hovem, Oscillations of polymeric microbubbles: Effect of the encapsulating shell. *Journal of the Acoustical Society of America*, 2000. **107**(4): p. 2272-2280.

[6] de Jong, N., et al., Absorption and scatter of encapsulated gas filled microspheres: Theoretical considerations and some measurements. *Ultrasonics*, 1992. **30**(2): p. 95-103.

[7] de Jong, N., R. Cornet, and C.T. Lancée, Higher harmonics of vibrating gas-filled microspheres. Part one: simulations. *Ultrasonics*, 1994. **32**(6): p. 447-453.

[8] de Jong, N., R. Cornet, and C.T. Lancée, Higher harmonics of vibrating gas-filled microspheres. Part two: measurements. *Ultrasonics*, 1994. **32**(6): p. 455-459.

[9] Glazman, R.E., Effects of adsorbed films on gas bubble radial oscillations. *The Journal of the Acoustical Society of America*, 1983. **74**(3): p. 980-986.

[10] Morgan, K.E., et al. *Experimental and theoretical analysis of individual contrast agent behavior*. in Ultrasonics Symposium, 1999. Proceedings. 1999 IEEE. 1999.

[11] Marmottant, P., et al., A model for large amplitude oscillations of coated bubbles accounting for buckling and rupture. *Journal of the Acoustical Society of America*, 2005. **118**(6): p. 3499-3505.

[12] Stride, E., The influence of surface adsorption on microbubble dynamics. *Philosophical Transactions of the Royal Society A-Mathematical Physical and Engineering Sciences*, 2008. **366**(1873): p. 2103-2115.
[13] King, L.V., On the Acoustic Radiation Pressure on Spheres. *Proceedings of the Royal Society of London. Series A, Mathematical and Physical Sciences*, 1934. **147**(861): p. 212-240.
[14] Watanabe, T. and Y. Kukita, Translational and Radial Motions of a Bubble in an Acoustic Standing-Wave Field. *Physics of Fluids a-Fluid Dynamics*, 1993. **5**(11): p. 2682-2688.
[15] Doinikov, A.A., Translational motion of a spherical bubble in an acoustic standing wave of high intensity. *Physics of Fluids*, 2002. **14**(4): p. 1420-1425.
[16] Hsiao, C.-T. and G. Chahine, Prediction of tip vortex cavitation inception using coupled spherical and nonspherical bubble models and Navier–Stokes computations. *Journal of Marine Science and Technology*, 2004. **8**(3): p. 99-108.
[17] Doinikov, A.A. and P.A. Dayton, Spatio-temporal dynamics of an encapsulated gas bubble in an ultrasound field. *J. Acoust. Soc. Am.*, 2006. **120**(2): p. 661.
[18] Dayton, P.A., J.S. Allen, and K.W. Ferrara, The magnitude of radiation force on ultrasound contrast agents. *The Journal of the Acoustical Society of America*, 2002. **112**(5): p. 2183-2192.
[19] O'Brien, J.P., N. Ovenden, and E. Stride, Accounting for the stability of microbubbles to multi-pulse excitation using a lipid-shedding model. *Journal of the Acoustical Society of America*, 2011. **130**(4): p. EL180-EL185.
[20] Becher, H.B., P., *Handbook of Echocardiography*2000, Frankfurt and New York: Springer Verlag.
[21] Gorce, J.-M., M. Arditi, and M. Schneider, Influence of Bubble Size Distribution on the Echogenicity of Ultrasound Contrast Agents: A Study of SonoVue(TM). *Investigative Radiology*, 2000. **35**(11): p. 661-671.

Towards the optimisation of acoustic fields for ablative therapies of tumours in the upper abdomen

P Gélat[1,2,†], G ter Haar[3] and N Saffari[2]

[1] National Physical Laboratory, Hampton Road, Teddington TW11 0LW, United Kingdom.

[2] Department of Mechanical Engineering, University College London, Torrington Place, London WC1E 7JE, United Kingdom.

[3] Therapeutic Ultrasound Group, Joint Physics Department, Institute of Cancer Research, Sutton SM2 5NG, United Kingdom.

[†] E-mail: Pierre.Gelat@npl.co.uk

Abstract. The efficacy of high intensity focused ultrasound (HIFU) for the non-invasive treatment of cancer has been demonstrated for a range of different cancers including those of the liver, kidney, prostate and breast. As a non-invasive focused therapy, HIFU offers considerable advantages over other techniques such as chemotherapy and surgical resection, in terms of its non-invasiveness and low risk of harmful side effects. There is, however, a number of significant challenges which currently hinder its widespread clinical application. One of these challenges is the need to transmit sufficient energy through the ribcage to induce tissue necrosis at the required foci whilst minimising the formation of side lobes and sparing healthy tissue. Ribs both absorb and reflect ultrasound strongly. As such, a common side effect of focusing ultrasound in regions located behind the rib cage is the overheating of bone and surrounding tissue, which can lead to skin burns. Successful treatment of a patient with tumours in the upper abdomen therefore requires a thorough understanding of the way acoustic and thermal energy are deposited. This is likely to rely on a treatment planning procedure in which optimal source velocity distributions are obtained so as to maximise a dose quantity at the treatment sites, whilst ensuring that this quantity does not exceed a specified threshold at other field locations, particularly on the surface of the ribs. Previously, a boundary element approach based on a Generalised Minimal Residual (GMRES) implementation of the Burton-Miller formulation was developed to predict the field of a multi-element HIFU array scattered by human ribs, the topology of which was obtained from CT scan data [1]. This work describes the reformulation of the boundary element equations as a least-squares minimisation problem with non-linear constraints. The methodology was subsequently tested at an excitation frequency of 100 kHz on a spherical multi-element array in the presence of a perfectly rigid cylindrical scatterer with hemi-spherical end-caps. Reduction of side lobes in the exterior domain and of acoustic pressure magnitudes on the surface of the cylinder were achieved whilst preserving a local maximum in the focal region.

1. Background

High intensity focused ultrasound (HIFU) enables highly localised, non-invasive tissue ablation and its efficacy has been demonstrated for treating a range of cancers, including abdominal tumours [2].

Content from this work may be used under the terms of the Creative Commons Attribution 3.0 licence. Any further distribution of this work must maintain attribution to the author(s) and the title of the work, journal citation and DOI.
Published under licence by IOP Publishing Ltd

There are nevertheless a number of significant challenges currently limiting its more widespread clinical application. In the cases of liver, kidney and pancreatic cancers, ribs may act as reflectors and aberrators. In addition to this, bone is a highly attenuating medium. As such, a common side effect of focusing ultrasound in regions located behind the rib cage is the overheating of bone and surrounding tissue, which can lead to skin burns over the location of the ribs [3], [4]. Hence, care must be taken so that sufficient energy is delivered through the rib cage to ablate tissue at the required location whilst also minimising the formation of side lobes. With the advent of piezoelectric multi-element transducer arrays used in conjunction with multi-channel electronics, there is an opportunity to address these challenges, as the magnitude and phase of each transducer element may be adjusted to create an optimal beam profile.

Botros *et al* [5] describe a method in which the design of a HIFU array was optimised using the pseudo-inverse technique (minimum norm least-squares solution) and by enforcing a constrained preconditioned pseudo-inverse method. The procedure calculates the required primary sources on the array while maintaining minimal power deposition over solid obstacles such as the ribs. Although the methodology endeavours to tackle the inverse problem described above, it suffers from the following limitation that the forward propagation model is 2D and relies on high frequency simplifications, and that the shape of the rib contours is idealised.

Liu *et al* [6] carried out a numerical study based on a modified Rayleigh-Sommerfeld integral approach where the feasibility of using a spherical ultrasound phased array for trans-rib liver tumour thermal ablation was investigated. Based on the feedback from anatomical imaging, array elements obstructed by the ribs were deactivated in an effort to minimise heat deposition on the ribs.

An approach based on adaptive optimisation has been discussed by Cauchard *et al* [7]. This was based on the analysis of signals backscattered from *ex vivo* ribs immersed in water, towards the HIFU array functioning in transmit/receive mode, in order to identify which elements required deactivation for minimisation of the reflected acoustic signal. Although this approach could be incorporated as part of a pre-operative procedure using a suitable forward model, it will not guarantee that the acoustic pressure on the surface of the ribs will not exceed a given threshold, since this condition is not a constraint.

Quesson *et al* [8] describe a method for selecting the elements of a HIFU transducer array to deactivate based on the relative location of the focal point and the ribs, identified from anatomical MR data. This method was implemented both *ex vivo* and *in vivo* in pig liver and was compared against the case in which all elements were activated. Temperature variations near the focus and ribs were monitored, demonstrating the benefit of deactivating selected array elements for sparing the ribs from excessive heating whilst still ensuring high enough temperatures for tissue ablation at the focus.

A similar approach was adopted by Marquet *et al* [9] where experiments investigating trans-rib HIFU using both *ex vivo* human ribs immersed in water and *in vivo* on pigs with 3D movement detection and compensation included.

The deactivation of transducer elements obstructed by ribs, or so-called binarised apodisation, described in [6], [7], [8] and [9], whilst practical in a clinical setting, is suboptimal, as acknowledged by Quesson *et al* [8]. This approach does not directly address the inverse problem of optimising the magnitudes and phases of the transducer element drive voltage to transmit sufficient energy through the ribcage to induce tissue necrosis at the required foci, whilst keeping the pressure on the ribs below a given threshold and ensuring minimal formation of side-lobes. Furthermore, binarised apodisation may hamper the treatment of deep-seated abdominal tumours in humans, where deactivating any number of elements may significantly reduce the ultrasonic intensities delivered at the focus.

Additionally, in order to reach the required temperature rise for tissue necrosis, an increase in treatment times may occur as a result of deactivating transducer elements. There is therefore a requirement to solve this inverse problem using a suitable forward model capable of addressing the effects of scattering and diffraction on 3D anatomical data. In the work described in this paper, the boundary element approach proposed by Gélat *et al* [1] was used, after implementation of dissipative mechanisms in the medium surrounding the ribs, in the form of a complex wave number. From the discretised form of the Helmholtz surface integral equation applied to locations on the surface of the scatterer and in the exterior domain, the inverse problem was formulated in which the unknowns were the complex velocities of each element on a transducer array.

Effects of nonlinear propagation were not considered as part of this work. It is however well known that HIFU fields can result in highly nonlinear behaviour in the focal region (Wu *et al*, 2004) leading to distortion of the acoustic waveform and transfer of energy from the fundamental frequency towards higher order harmonics. From the point of view of treatment planning, it has been suggested that correcting for aberrations introduced by the presence of bone need not require nonlinear propagation models and that linear approaches may generally suffice [10]. Furthermore, nonlinear behaviour is likely to be mainly confined to the central focal lobe. Hence, although a more rigorous treatment of the focal region may be required, useful information can be obtained from linear models when investigating energy depositions at the surface of the ribs.

The reformulation of the boundary element approach was tested on a reduced-complexity problem involving a 256 element spherical phased array with pseudo-random distribution of the elements on its surface. Each element was vibrating at a frequency of 100 kHz, and a perfectly rigid cylindrical scatterer was placed between the array and its geometric focus. Using a Numerical Algorithms Group (NAG) Numerical Library [11] routine which calculates the minimum of a sum of squares for a set of nonlinear constraints, optimal values for the real and imaginary parts of the element velocities were obtained so that the acoustic pressures at specified locations in the exterior domain best fitted a required field distribution in a least-squares sense. Implementation of the constraints ensured that the acoustic pressure magnitudes on the surface of the cylinder did not exceed a specified threshold and that the element velocity magnitudes were kept within the specified dynamic range.

2. Theory

Consider an exterior domain V_{ext} bounded by a surface Λ and by a closed smooth surface S. Let Λ be at a sufficiently large distance from the acoustic sources and from the surface S. Let the boundary condition on Λ satisfy Sommerfeld's acoustic radiation condition. The propagation of time-harmonic acoustic waves in a homogeneous isotropic inviscid medium is described by the Helmholtz equation:

$$\nabla^2 p(\vec{r}) + k^{*2} p(\vec{r}) = 0, \forall \vec{r} \in V_{ext} \tag{1}$$

where p is the acoustic pressure, $k^* = \omega / c^*$ is the acoustic wave number where ω is the angular frequency, c^* the complex sound speed in the external medium and $\vec{r}$ is the position vector. A time harmonic dependence is implicitly assumed, so that the acoustic pressure is given by the real part of $p(\vec{r})e^{i\omega t}$. A source term may be included in the right hand side of (1) (incident pressure wave), but it is convenient to split the total pressure $p(\vec{r})$ as the sum of the incident pressure $p_i(\vec{r})$ and the pressure scattered by the surface S, $\forall \vec{r} \in V_{ext}$.

For scattering problems, the integral representation of the solution to (1) is given by the Helmholtz integral equation:

Journal of Physics: Conference Series **457** (2013) 012002 doi:10.1088/1742-6596/457/1/012002

$$\int_S \left[p(\vec{r}_q)\frac{\partial G(\vec{r}|\vec{r}_q)}{\partial n_q} - \frac{\partial p(\vec{r}_q)}{\partial n_q}G(\vec{r}|\vec{r}_q)\right]dS = \frac{1}{2}p(\vec{r}) - p_i(\vec{r}), \forall\vec{r} \in S \tag{2}$$

$$\int_S \left[p(\vec{r}_q)\frac{\partial G(\vec{r}|\vec{r}_q)}{\partial n_q} - \frac{\partial p(\vec{r}_q)}{\partial n_q}G(\vec{r}|\vec{r}_q)\right]dS = p(\vec{r}) - p_i(\vec{r}), \forall\vec{r} \in V_{ext} \tag{3}$$

where the Green's function in a three-dimensional space is given by:

$$G(\vec{r}|\vec{r}_q) = \frac{1}{4\pi\|\vec{r}-\vec{r}_q\|}e^{-ik^*\|\vec{r}-\vec{r}_q\|} \tag{4}$$

and $p(\vec{r}_q)$ is the pressure on the surface S at point $\vec{r}_q$ and $\vec{n}_q$ is the outward normal.

The problem described by the integral equation (2) suffers from non-uniqueness at frequencies of excitation approaching an eigenvalue of one of the (fictitious) modes of the cavity inside the scatterer. In cases where the dimensions of the scatterer are large compared with the wavelength in the propagating medium, such as in HIFU applications, this is likely to occur. The matrix formed by discretising equation (2) then becomes close to singular. When the wave number is purely real, the method which appears to offer the best compromise in terms of application of the Helmholtz integral equation to exterior acoustic problems involving scatterers of arbitrary shape remains the Burton-Miller formulation [12], which solves for a linear combination of equation (2) and its derivative with respect to the outward normal vector on S at $\vec{r}$. However, for a wave number whose imaginary part is non-zero, solving the discretised form of equation (2) may be sufficient, as the fictitious modes of the cavity inside the scatterer are likely to be dampened out. For the problem investigated in this paper, it will be demonstrated in Section 3 that a discretisation of the Helmholtz integral equation is indeed sufficient and that the coupling term involving the normal derivative formulation need not be considered. The methodology described below for solving the inverse problem may nevertheless be generalised to the Burton-Miller formulation.

By discretising equation (2), for all position vectors $\vec{r}$ on S corresponding to each node on the mesh of the surface, a linear system of equations may be generated. For a perfectly rigid scatterer, the normal derivative terms in equation (2) are zero.

$$[H]\{p_{surf}\} = -\{p_i\} \tag{5}$$

where the acoustic pressures on the surface S have been relabelled $\{p_{surf}\}$.

Let the incident field be a linear combination of plane circular pistons. The incident field on S is a linear combination of the source velocities of each piston. Equation (5) may therefore be re-written as:

$$[H]\{p_{surf}\} = -[\beta]\{U\} \tag{6}$$

where the elements of $[\beta]$ may be obtained analytically in the far-field or by solving the Rayleigh integral in the near-field. $[\beta]$ is of dimension $M\times N$, where M is the number of nodes on the surface S after its discretisation and N is the number of plane pistons.

Consider equation (3), where the position vectors are located in the exterior volume V_{ext}. Again, the term on the left hand side involving the normal derivative of the acoustic pressure on S is zero for a perfectly rigid surface. When evaluated numerically, the integral may be expressed as a weighted sum of the pressures on S. Additionally, the incident pressure at any given field location is a linear combination of the source velocities. Hence, for a specified number of locations in the exterior volume, we have:

$$[Q]\{p_{surf}\} = \{p_{ext}\} - [\gamma]\{U\} \tag{7}$$

where the acoustic pressures in the exterior volume V_{ext} have been relabelled $\{p_{ext}\}$. The coefficients in $[\gamma]$ may be obtained from the Rayleigh integrals relating each plane piston on the multi-element array to each location in V_{ext}.

The vector of surface pressures may be eliminated by combining equations (6) and (7).

$$([\gamma] - [Q][H]^{-1}[\beta])\{U\} = \{p_{ext}\} \tag{8}$$

Equation (6) may be re-written as follows.

$$-[H]^{-1}[\beta]\{U\} = \{p_{surf}\} \tag{9}$$

or

$$[A]\{U\} = \{p_{surf}\} \tag{10}$$

Equations (8) may be re-written as follows.

$$[C]\{U\} = \{p_{ext}\} \tag{11}$$

All quantities in equation (11) will generally be complex. When investigating an inverse problem, it is often convenient to deal with an objective function of real variables. As such, equation (11) is rewritten as follows,

$$\begin{bmatrix} \mathrm{Re}[C] & -\mathrm{Im}[C] \\ \mathrm{Im}[C] & \mathrm{Re}[C] \end{bmatrix} \begin{Bmatrix} \mathrm{Re}\{U\} \\ \mathrm{Im}\{U\} \end{Bmatrix} = \begin{Bmatrix} \mathrm{Re}\{p_{ext}\} \\ \mathrm{Im}\{p_{ext}\} \end{Bmatrix} \tag{12}$$

or

$$[\hat{C}]\{\hat{U}\} = \{\hat{p}_{ext}\} \tag{13}$$

We wish to obtain a set of real and imaginary parts of source velocities which best fit a prescribed field pressure distribution in a least-squares sense such that:

- the acoustic pressure magnitudes on S do not exceed a threshold defined by $p_{surf_{max}}$
- the source velocity magnitudes do not exceed the upper bound of the dynamic range of each element U_{max}.

This may be expressed as a least-squares minimisation problem with nonlinear constraints:

$$\min_{\hat{U}} \frac{1}{2}\left\|[\hat{C}]\{\hat{U}\} - \{\hat{p}_{ext}\}\right\|_2^2 \text{ such that } \begin{cases} \left|\{p_{surf}\}\right| \leq p_{surf\,max} \\ |U| \leq U_{max} \end{cases} \tag{14}$$

The HIFU transducer modelled as part of the underlying work was assumed to be a spherically shaped bowl and populated with N = 256 plane circular elements mounted onto its surface. The elements were each of a = 1 cm radius. A radius of curvature of D = 50 cm was used. The outer diameter of the HIFU transducer was chosen as 60 cm. The elements were pseudo-randomly spatially distributed on the surface of the array. The excitation frequency was assumed to be 100 kHz. A frontal view of the array is shown in figure 1.

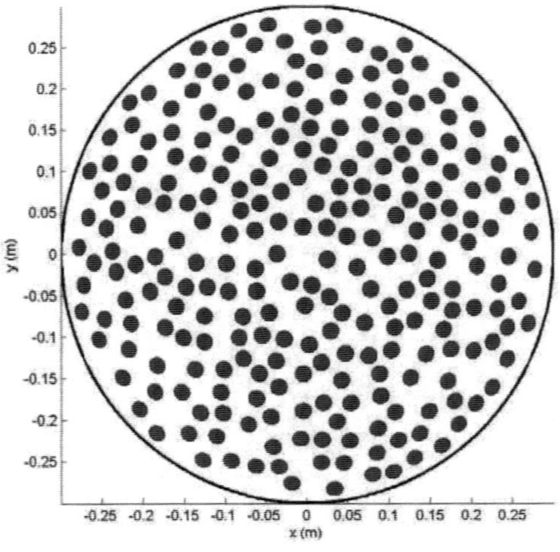

Figure 1. Frontal view of HIFU random phased array configuration. 1 cm element diameter, 60 cm array diameter, 50 cm focal length, 100 kHz frequency of operation.

It is acknowledged that both the frequency of operation of the array and its dimensions do not conform to what is commonly used for transcostal HIFU [13]. Investigating a reduced-size problem such as this one is nevertheless both relevant and beneficial prior to applying the methodology to transcostal HIFU frequencies and dimensions, which will involve much greater computational resources.

3. Results

The Helmholtz integral equation boundary element approach was implemented on a dedicated computer cluster. The approach used here differs from that described in [1] as follows.

- The coupling coefficient in the Burton-Miller (normal derivative) formulation was set to zero.
- The speed of sound in the exterior was specified as a complex number.
- Operations were vectorised to solve equation (5) for multiple right hand sides simultaneously, thus enabling the coefficients of $[A]$ to be obtained more efficiently.

The routines employed to generate the boundary element matrices were obtained from the PAFEC (Program for Automatic Finite Element Calculations) VibroAcoustics software with permission from PACSYS Ltd.

A validation on a simple scatterer was carried out to justify setting the coupling coefficient in the Burton-Miller formulation to zero. The results displayed here are for a 100 kHz unit amplitude plane wave scattered by a perfectly rigid cylinder with hemi-spherical end-caps. The cylinder was 22 cm in length and 1 cm in diameter. These were compared against the analytical solution for a cylinder of infinite length provided in [14] in the shadow zone between 1.2 cm and 50.0 cm (see figure 2).

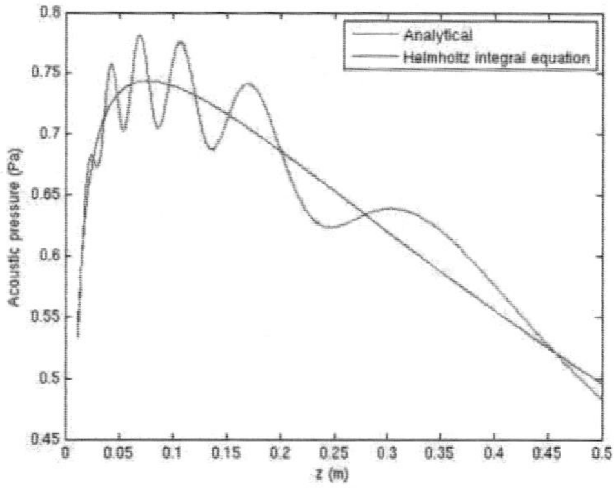

Figure 2. Helmholtz integral equation implementation of BE formulation on a cylindrical scatterer of 2 cm diameter and 22 cm length, with hemispherical end-caps. Incident field: unit amplitude 100 kHz plane wave travelling in positive z direction. Comparison against analytical solution involving infinite cylinder.

The scatterer was meshed using isoparametric eight-noded quadratic patches ensuring at least three elements per wavelength for a wavespeed of 1500+4.405i m s^{-1}, corresponding to an absorption coefficient of 47.2 Np per metre at 1 MHz (assuming a power law with a linear dependence on frequency). The density of the medium was assumed to be 1000 kg m^{-3}. Figure 2 shows agreement within 1.9% of the analytical solution in the shadow zone, at 0.2 cm from the surface of the cylinder along the z-axis. Oscillations about the analytical solution are observed, which are due to constructive and destructive interference from the waves diffracted by the end-caps of the cylinder.

After completing this validation, the incident field was replaced with that of the array shown in figure 1. A simulation solving equation (4) for multiple right hand sides was launched so that the product $[H]^{-1}[\beta]$ could be obtained in a single run. The axis of symmetry of the cylindrical scatterer was chosen to be parallel to the x-axis and located -5 cm away from the geometric focus of the array in the z-direction (towards the array). This arrangement is illustrated in figure 3.

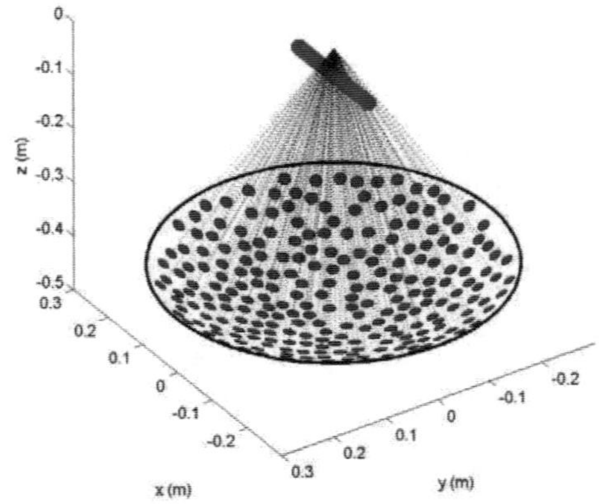

Figure 3. Position of cylindrical scatterer with respect to focused array.

From a knowledge of $[H]^{-1}[\beta]$, the acoustic pressure on the surface of the cylinder may be obtained using equation (6) for a given distribution of source velocities. In order to formulate the objective function and the constraints, the case of all elements vibrating in phase and with unit velocity amplitude was investigated. The acoustic pressure magnitudes at selected locations in the *y*-*z* plane are shown in figure 4. The corresponding incident pressure field (i.e. in absence of the scatterer) is shown in figure 5.

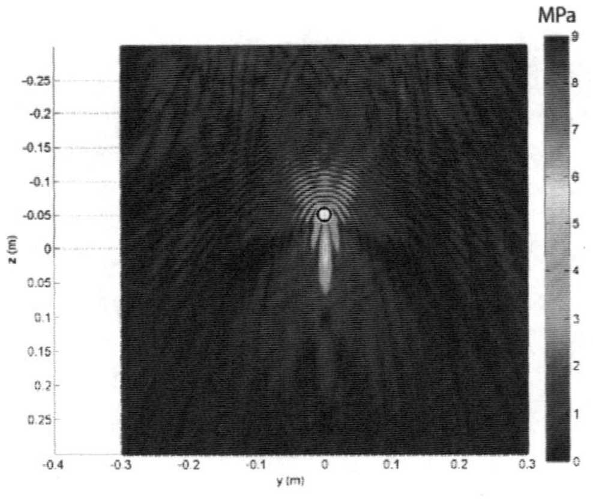

Figure 4. Acoustic pressure magnitude in *y*-*z* plane resulting from field of 100 kHz multi-element array. Uniform unit amplitude velocity and zero phase. Contour of cylinder shown in black.

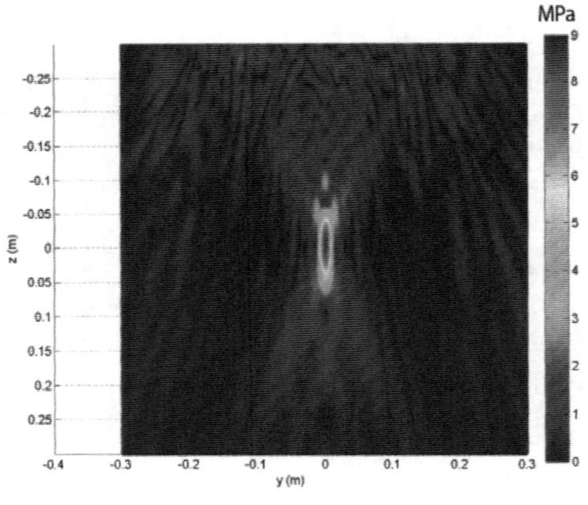

Figure 5. Incident acoustic pressure magnitude in *y*-*z* plane resulting from field of 100 kHz multi-element array. Uniform unit amplitude velocity and zero phase.

The resulting acoustic pressure magnitude on the surface of the cylinder is shown in figure 6.

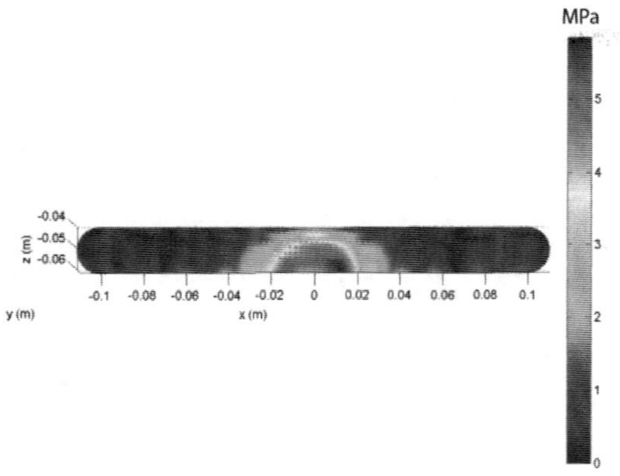

Figure 6. Acoustic pressure magnitude on surface of cylinder resulting from field of 100 kHz multi-element array. Uniform unit amplitude velocity and zero phase.

Figures 4 and 5 shows that inserting the cylindrical scatterer between the array and the focus causes the acoustic pressure magnitude at the focus to drop from 9 MPa to 6 MPa. Figure 6 shows that the maximum pressure amplitude on the surface of the cylinder is 6 MPa for all elements of the array vibrating at an amplitude of 1 m s^{-1} and with uniform phase. Based on these results, the value $p_{surf_{max}}$ in equation (14) was chosen as 4 MPa. U_{max} was taken as 1 m s^{-1}. The vector of pressures in the exterior volume in the objective function (i.e. the 'desired' field pressure distribution) was generated from incident pressure field values at 5192 equally spaced locations in the *y-z* plane such that -3 cm $\leq y \leq$ 3 cm and -3 cm $\leq z \leq$ 3 cm, the focus of the array being at the global origin. Locations inside a 1.5 cm radius around the axis of the scatterer were removed.

The NAG Numerical Library e04us routine was used to carry out the constrained minimisation [11]. It is designed to minimise an arbitrary smooth sum of squares function subject to constraints, which may include simple bounds on the variables, linear constraints and smooth nonlinear constraints. It employs a sequential quadratic programming method [15]. The partial derivatives of the constraints and of the objective function with respect to the minimisation variables were supplied. According to [11], scaling of the problem is likely to reduce the number of iterations required and make the problem less sensitive to perturbations in the data, thus improving the condition of the problem. It is suggested that, in the absence of better information it is sensible to make the Euclidean lengths of each constraint of comparable magnitude. The problem was therefore scaled by a factor of 10^{-6} and the constraints involving the source velocities were scaled by a factor of 10^{6} so that they were of the same order of magnitude as the surface pressure magnitude constraints. Initial values of the minimisation variables were all specified as $1/\sqrt{2}$ m s^{-1} so that the magnitudes of the source velocities were initially at the upper end of the specified dynamic range. This generated a solution for the optimisation variables where both sets of constraints were satisfied. Were the rescaling not carried out prior to the minimisation, the NAG solver either became unstable or returned results which did not satisfy the constraints.

The real and imaginary parts of the velocities were subsequently rescaled and the surface and field pressures calculated. Figures 7 and 8 respectively display the magnitudes and phases of the element velocities of the array resulting from the constrained minimisation. It was verified that the velocity magnitudes did not exceed 1 m s^{-1}.

Journal of Physics: Conference Series **457** (2013) 012002 doi:10.1088/1742-6596/457/1/012002

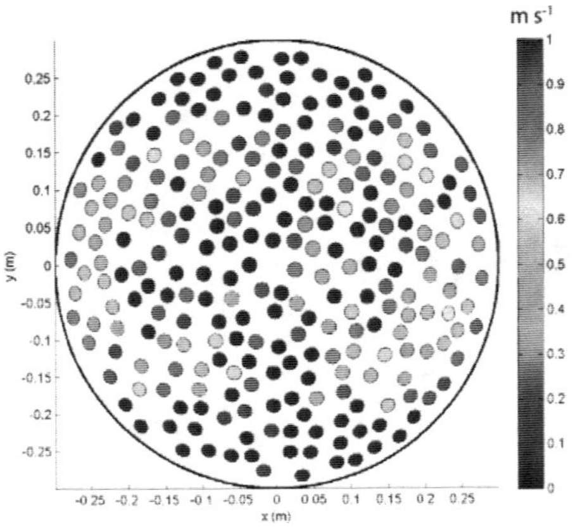

Figure 7. Source velocity magnitudes resulting from constrained minimisation.

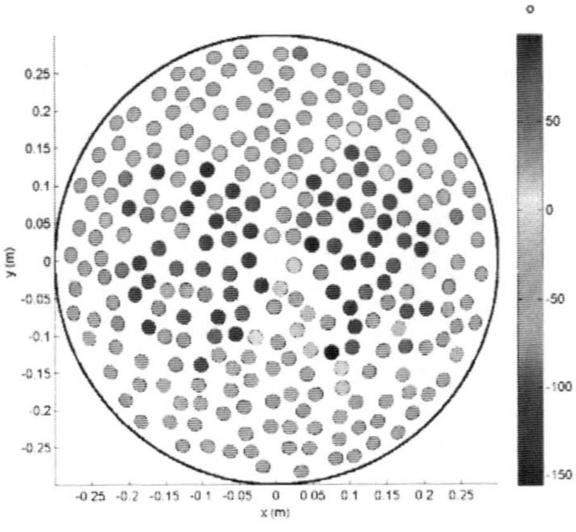

Figure 8. Source velocity phases resulting from constrained minimisation.

Figure 9 shows the acoustic pressure magnitude on the surface of the cylinder, resulting from the source velocity distribution displayed in figures 7 and 8, which was verified not to exceed 4 MPa.

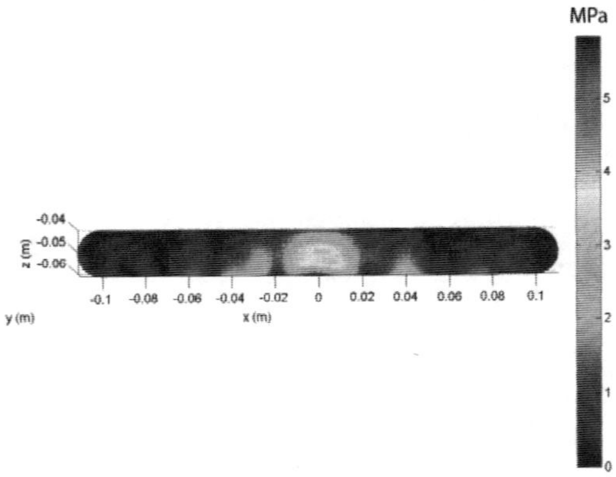

Figure 9. Acoustic pressure magnitude on surface of cylinder resulting from field of 100 kHz multi-element array. Source velocity distribution obtained from constrained minimisation.

Figure 10 shows the acoustic pressure magnitude in the *y*-*z* plane resulting from the source velocity distribution displayed in figures 7 and 8.

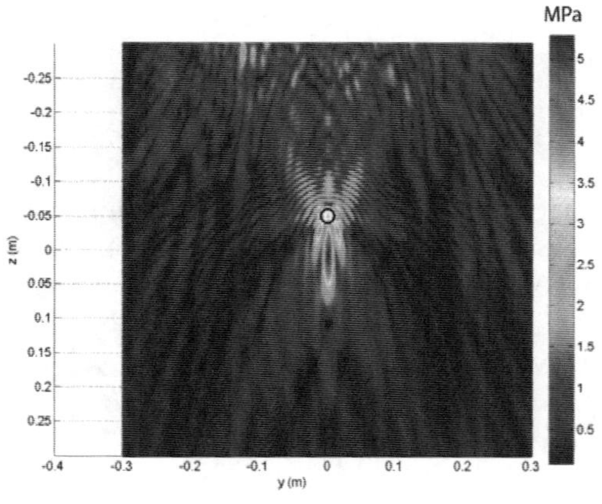

Figure 10. Acoustic pressure magnitude in *y-z* plane resulting from field of 100 kHz multi-element array. Source velocity distribution obtained from constrained minimisation.

This particular configuration makes it challenging for the majority of the acoustic energy to be transmitted in the vicinity of the focus, owing to the close proximity of the scatterer to the focal region of the array. This is further hindered by the cylinder's large diameter compared with the beam of the array, as visualised in figure 3. The acoustic pressure magnitude at the focus shown in figure 10 is reduced by 14% compared with the case where all elements are vibrating in phase and with unit velocity amplitude. To investigate how side-lobes were affected by the constrained minimisation, the acoustic pressure magnitudes normalised to their value at the focus are displayed in figure 11, where all elements are vibrating with unit amplitude and uniform phase, and in figure 12, where the optimised velocity values are used..

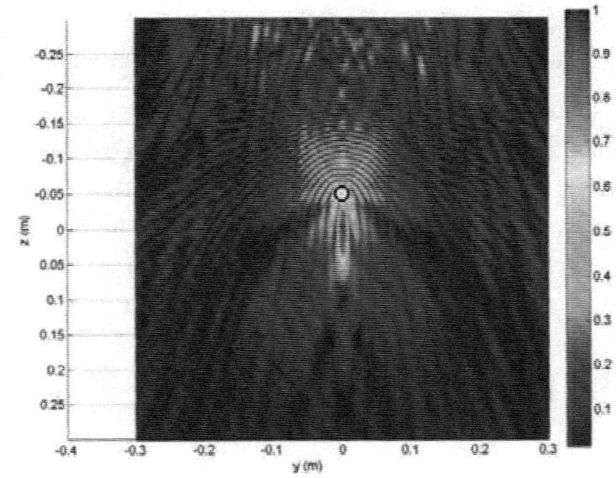

Figure 11. Normalised acoustic pressure magnitude in *y-z* plane resulting from field of 100 kHz multi-element array. Uniform unit amplitude velocity and zero phase.

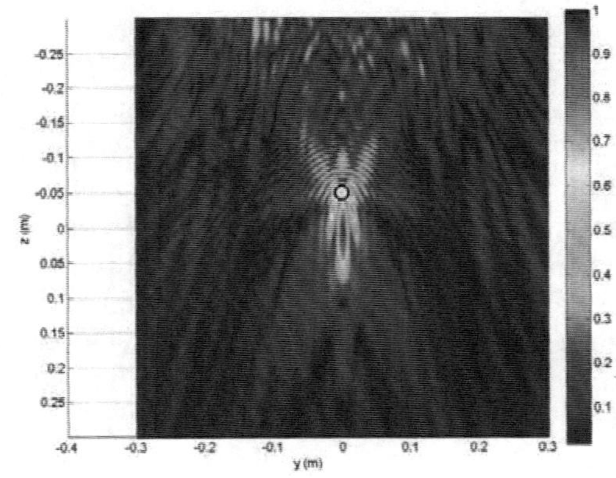

Figure 12. Normalised acoustic pressure magnitude in *y-z* plane resulting from field of 100 kHz multi-element array. Source velocity distribution obtained from constrained minimisation.

Figure 12 shows that a qualitative overall reduction in the side-lobes relative to the focus was achieved as a consequence of the constrained minimisation compared with the uniform amplitude and phase results displayed in figure 11.

4. Conclusions

An approach based on a GMRES implementation of the Helmholtz integral equation boundary element formulation was validated to model the scattering of the field of a multi-element random HIFU phased array by 3D objects of an arbitrary geometry under continuous wave excitation. The medium surrounding the scatterer featured a complex speed of sound thus allowing a phenomenological approach to modelling attenuation effects.

Using the discretised form of the Helmholtz integral equation for locations in the exterior volume and on the surface of the scatterer, the inverse problem of determining the complex velocities of a multi-element array which produces an acoustic pressure field that best fits a required acoustic pressure distribution in a least-squares sense was formulated such that:

- The pressure magnitude on the surface of the scatterer did not exceed a specified threshold.
- The amplitude of the velocity of each element on the array was bounded by maximum value defined by the dynamic range.

The approach was tested on a reduced-size problem involving a 100 kHz 256-element spherical array in presence of a cylindrical scatterer. Employing a NAG library solver, a least-squares minimisation with nonlinear constraints was carried out to solve the above problem, where the gradients of the objective function and of the constraints with respect to the optimisation variables were provided. The solver returned a set of real and imaginary parts for the source velocities which satisfied both sets of constraints, hence reducing side lobes and acoustic pressures on the surface of the scatterer compared with the case of all elements being driven with uniform phase and amplitude.

This technique shows promise in terms of providing an optimal solution to the transcostal HIFU problem of sparing the ribs from excessive ultrasound-induced heating whilst ensuring sufficient energy deposition at the treatment location and minimising side-lobes. Furthermore, the optimal velocity magnitude distribution suggests that only two to three elements come close to being deactivated, thus suggesting that biniarised apodisation may be suboptimal. Simulations on human anatomical ribs at 1 MHz are currently being carried out and will form the basis of another paper. As part of this work, the scatterer was assumed to be perfectly rigid. In reality, it is known that both longitudinal and shear waves can be generated in bone [16], [17]. As bone is a highly attenuating medium compared to soft tissue, absorption of shear waves may play an important part of heating in the bone. A full elastodynamic boundary element formulation would be required to deal with this phenomenon accurately. It is acknowledged that this is a limitation of the method proposed in this paper. Nevertheless, a relatively simple extension of the proposed formulation could account for a locally reacting impedance boundary condition on the surface of the scatterer, which would represent a step beyond assuming that the ribs are acoustically hard surfaces.

Acknowledgements

The authors would like to acknowledge the support of the National Measurement Systems Directorate of the UK Department of Business, Innovation and Skills and EPSRC EP/F025750/1. The authors are grateful to PACSYS Ltd. for supplying stand-alone executables of the Program for Automatic Finite Element Calculations boundary element routines as well as for useful discussions. The authors would equally like thank Minh Hoang for advice on inverse methods and help with NAG Numerical Library solvers. The authors would like to express their gratitude to David Sinden, Simon Arridge and Timo Betcke for useful discussions.

References

[1] Gélat P, ter Haar G and Saffari N 2011 Modelling of the acoustic field of a multi-element HIFU array scatterred by human ribs *Phys. Med. Biol.* **56** 5553.

[2] Crum L, Bailey M, Hwang J H, Khokhlova V and Sapozhnikov 2010 Therapeutic ultrasound: Recent trends and future perspectives *Physics Procedia* **3** 25

[3] Wu F, Zhi-Biao W, Wen-Zhi C, Hui Z, Jin B, Jian-Zhong Z, Ke-Quan L, Cheng-Bing J, Fang-Lin X, Hai-Bing S 2004 Extracorporeal high intensity focused ultrasound ablation in the treatment of patients with large hepatocellular carcinoma *Ann. Surg. Oncol.* **11** 1061

[4] Li J-L, X-Zh Liu, Zhang D, Gong X-F 2007 Influence of ribs on nonlinear sound field of therapeutic ultrasound. *Ultrasound Med. Biol.* **33** 1413

[5] Botros Y Y, Ebbini E S, Volakis J L 1998 Two-step hybrid virtual array-ray (VAR) technique for focusing through the ribcage *IEEE Trans. Ultrason. Ferroelectr. Freq. Control* **45** 989

[6] Liu H-L, Chang Hsu, Chen W-S, Shih T-C, Hsiao J-K and Lin W-L. Feasibility of transrib focused ultrasound thermal ablation for liver tumours using a spherically curved 2D array: A numerical study 2007 *Med. Phys.* **34** 3436

[7] Cochard E, Prada C, Aubry J F, Fink M. Ultrasonic focusing through the ribs using the DORT method 2009 *Med. Phys.* **36** 3495

[8] Quesson B, Merle M, Roujol S, de Senneville B D and Moonen C T. A method for MRI guidance of intercostal high intensity focused ultrasound ablation in the liver 2010 *Med. Phys.* **37** 2533

[9] Marquet F, Aubry J F, Pernot M, Fink M and M Tanter. Optimal transcostal high-intensity focused ultrasound with combined real-time 3D movement tracking and correction 2011 *Phys. Med. Biol.* **56** 7061

[10] Pinton G, Aubry J-F, Fink M and Tanter M 2011 Effects of nonlinear ultrasound propagation on high intensity brain therapy *Med. Phys.* **38** 1207

[11] http://www.nag.co.uk/numeric/numerical_libraries.asp

[12] Burton A J and Miller G F 1971 The application of integral equation methods to the numerical solution of some exterior boundary-value problems *Proc. R. Soc. London Ser.* **A323** 201

[13] Gavrilov L R, Hand J W 2000 A theoretical assessment of the relative performance of spherical phased arrays for ultrasound surgery *IEEE Trans. Ultrason. Ferroelectr. Freq. Control* **47** 125

[14] Morse P M and Ingard K U, *Theoretical Acoustics* 1968 New York. CO: McGraw-Hill 401

[15] Bonnans J F, Gilbert J C, Lemaréchal C, Sagastizábal C A, *Numerical optimization: Theoretical and practical aspects* 2006 Berlin: Springer-Verlag 490

[16] Kohles S S, Bowers J R ,Vailas A C and Vanderby, R 1997 Ultrasonic wave velocity measurement in small polymeric and cortical bone specimens J. Biomech. Eng. **119**, 232

[17] Mast D, Hinkelman L, Metlay L, Orr M and Waag R 1999 Simulation of acoustic pulse propagation and attenuation in the human chest wall *J. Acoust. Soc. Am.***106** 3665

Journal of Physics: Conference Series **457** (2013) 012003 doi:10.1088/1742-6596/457/1/012003

Characterisation of ultrasonic structural noise in multiple scattering media using phased arrays

T Bedetti[1], V Dorval[1], F Jenson[1] and A Derode[2]

[1] CEA, LIST, F-91191, Gif-sur-Yvette cedex, France

[2] Institut Langevin, Université Paris Diderot – Paris 7, ESPCI ParisTech, CNRS UMR 7587, 10 rue Vauquelin, 75005 Paris, France

E-mail: thomas.bedetti@cea.fr, vincent.dorval@cea.fr

Abstract. The ultrasonic inspection of multiple scattering media gives rise to structural noise which makes it difficult to detect potential damage or crack inside the component. In order to predict the performances of ultrasonic inspection over such complex media, scattering models can be used. Such models rely on specific key parameters describing the multiple scattering process, which can be determined by specific measurements and post-processing techniques. Such experiments were carried out on stainless steel plates using linear phased-arrays. They consist in recording the response matrix constituted by impulse responses between all the elements of the array. By conducting post-processing on this matrix, we measure the elastic mean free path l_e and the correlation distance d_c of the recorded noise. Additionally, the dynamic behaviour of the coherent backscattering effect was studied in order to measure the diffusion constant D. Plane-wave beamforming has been applied to the response matrix to improve the angular resolution and the signal-to-noise ratio of the backscattered intensity. Details of post-processing techniques will be shown.

1. Introduction

In the context of Non Destructive Testing (NDT), the acoustical inspection of scattering media represents a scientific challenge. Such materials are widely used in the industry, from stainless steel pipes used in nuclear power plants to concretes and composite materials used respectively in construction and aeronautical industries. Due to their complex microstructure, they are difficult to inspect. Indeed, the interaction of an ultrasonic wave with the microstructure leads to the deformation of the incident beam by scattering. In the case of stainless steels, the grain boundaries act as an impedance contrast. When the wavelength is of the same order of magnitude as the size of these grains, the wave is scattered and its energy is spread over the medium, attenuating the incident beam and making a "structural noise" appear. The amplitude of this noise can even become higher than echoes corresponding to the presence of defects in the structure. In this case, the quality of the inspection is compromised. Two main phenomena contribute to the noise: single scattering, when only one scatterer has been involved before the recording of the wave, and multiple scattering, when the wave is scattered several times along its path. There is no clear limit between these two regimes but it is sometimes possible to neglect one of the contributions in comparison to the other.

Simulations are often used to evaluate the detection capabilities of inspection methods. For this purpose, CIVA, the NDT simulation software developed at the CEA, includes different models that simulate the propagation of ultrasonic waves in various media. In the case of polycrystalline structure,

Content from this work may be used under the terms of the Creative Commons Attribution 3.0 licence. Any further distribution of this work must maintain attribution to the author(s) and the title of the work, journal citation and DOI.
Published under licence by IOP Publishing Ltd

a noise generator based on the single scattering hypothesis has been developed [1]. Nevertheless, when multiple scattering effects cannot be neglected, the noise level is underestimated. A new noise simulation method is currently developed to take these effects into account. Contrary to numerical approaches such as finite element methods [2], it uses a semi analytical approach to model multiple scattering. It requires parameters that characterize the scattering as input.

The propagation of ultrasonic waves in the multiple scattering regime has been widely studied. Turner and Weaver proposed a study of the incoherent intensity propagation in polycrystalline materials based on the Radiative Transfer Equation [3]. Also, Page tested the diffusion approximation to describe the propagation trough a random medium [4]. In these models, it is common to characterise the transport properties in a disordered medium in terms of the elastic mean free path l_e, the transport mean free path l^* and the diffusion constant D. In addition, Derode studied the statistics of speckle patterns obtained by recording structural noise [5]. Though this study was limited to single scattering in anisotropic composites, Derode showed that the correlation distance of the noise d_c was influenced by the microstructure. The new simulation method takes these three parameters as entry values.

However, the experimental determination of l_e , D and d_c classically needs different setups. In this paper, we show that it is possible to estimate the three parameters via one single recording. In NDT, N-element phased arrays are more and more used to inspect structures. A widely used process is to perform the so-called Full Matrix Capture (FMC) to record the response matrix K of the array, which contains all the impulse responses between every element of the array. By applying post-processing techniques to K, it is possible to extract various data. The study of the coherent field can lead to l_e, of the incoherent field to d_c, and of the intensity to D by highlighting the coherent backscattering effect.

2. Experimental setup and post-processing techniques

In the experiment, the transducer and the sample are placed in a water tank. We used two N-element ultrasonic arrays whose characteristics are described in table 1. The sampling frequency is 40MHz.

The experiment consists in measuring the response matrix K, i.e. the interelement matrix of the array (see figure 1). A 2-period-long sine is emitted from transducer e into the scattering sample. The backscattered wave is recorded with the N transducers of the same array. The operation is repeated for the N emitting transducers. The NxN array response matrix $K(t)$ whose elements are the N^2 impulse responses $k_{er}(t)$ is finally obtained. The entire measure is then repeated for P positions of the array, in order to perform averaging over the distribution of the scatterers into the medium.

The recording of K is done in two forged stainless steel plates. These plates have the same thickness $d = 24$mm and differ in the size of their grains γ, 130μm and 500μm. From here on, they will be called, respectively, SG (Small Grains) and CG (Coarse Grains) samples. The elastic properties of the plates can be found in table 2. The distance between the sample and the array is $h = 100$mm. The number of positions of the array differs according to the sizes of the sample. 55 matrices are recorded in the CG sample and 90 in the SG sample.

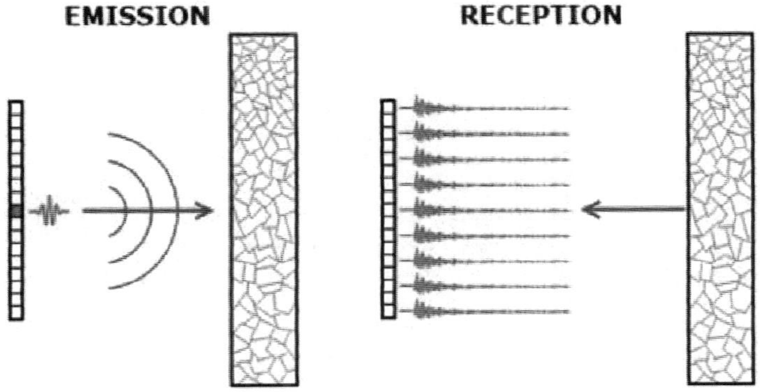

Figure 1. Experimental setup.

Table 1. Characteristics of the linear phased arrays.

Reference	Central frequency	-6dB Bandwidth	N	Element size	Pitch	Array size
SONO12	1.5 MHz	1.2 – 1.9 MHz	64	0.5 x 12 mm	0.55 mm	35.15 mm
SONO65	3 MHz	2 – 3.8 MHz	64	0.389 x 12 mm	0.417 mm	26.66 mm

Table 2. Elastic properties of the forged stainless steel plates.

v_L	v_T	ρ	C_{11}	C_{12}	C_{44}
5650 m.s^{-1}	3100 m.s^{-1}	8.12 g.cm^{-3}	169 GPa	145 GPa	123 GPa

In order to obtain the searched parameters, we chose to perform beamforming to the response matrix, as is usually done with ultrasonic arrays for instance in medical imaging [6]. By applying delay laws along the dimensions of K, it is possible to simulate various experimental setups:

- Far field conditions using plane waves.
- Near field conditions using virtual point source.
- Focused transducers using cylindrical delay laws.

In this paper, we will use plane wave beamforming. First, we introduce a linear delay between impulse responses $k_{er}(t)$ so that signals from the required directions are brought into phase. Second, the time-shifted signals are added. At the cost of longer computation times than direct measurements, the signal to noise ratio is higher due to the fact that every impulse response contributes to the final signal $s_{\alpha\beta}(t)$, with α and β the angles of interest respectively in emission and reception.

Without additional correction, and due to the finite size of the array, edge effects can disturb the precision of the beamforming. To correct it, apodisation is performed with the help of a Hann window to assure the planarity of the emitted wave.

3. Determination of the parameters

3.1. *Coherent field and elastic mean free path*

During the propagation of an ultrasonic wave in a polycrystalline medium, each scattering that occurs leads to the spreading of a part of the energy over the sample. In consequence, the intensity of the coherent wave that resists to averaging is progressively attenuated. The characteristic distance of this attenuation is called the elastic mean free path, l_e. It is assumed that attenuation by intrinsic absorption is negligible compared to attenuation by scattering.

Knowing the value of l_e and the thickness d of the plate, the ratio l_e/d is a useful information about the scattering regime inside the medium [7]. If it is above one, the coherent wave is still dominant and multiple scattering can be neglected. On the contrary, if it is less than one, the coherent wave lost the majority of its energy by the time it reaches the back-surface of the plate. In that case, single scattering becomes negligible and multiple scattering dominates. This regime is often called the "diffusion regime" due to the way the acoustic energy is propagating into the medium. More information will be found in paragraph 3.3.

Classically, the measure of l_e is done using a transmission setup, working on samples of different thicknesses. Here, as we impose the use of the response matrix obtained in reflection, the calculation method is based on the study of the two first back-wall echoes spectra [8]. Taking into account absorption, attenuation, and the interface between the couplant and the medium, we can write the amplitude of the two echoes spectra of a plane wave emitted at zero incidence as a function of the incident wave $|A_0(\omega)|$.

$$|A_{echo\,1}(\omega)| = |A_0(\omega)|\, e^{-2\alpha_1(\omega)h}\, e^{-2\alpha_2(\omega)d}\, T_{12}R_{22}T_{21} \quad (1)$$

$$|A_{echo\,2}(\omega)| = |A_0(\omega)|\, e^{-2\alpha_1(\omega)h}\, e^{-4\alpha_2(\omega)d}\, T_{12}R_{22}^3T_{21} \quad (2)$$

where α is the attenuation coefficients, T and R are the transmission and reflection coefficients, and the subscripts 1 and 2 correspond respectively to the coupling medium and the sample, as the back-wall medium is the same as the couplant. Dividing one expression by the other, we obtain a realisation of the attenuation function of the coherent field, $\alpha_2(\omega)$. As the elastic mean free path is the characteristic distance of attenuation of the intensity, it is proportional to the inverse of the attenuation function:

$$l_e(\omega) = \frac{1}{2\alpha_2(\omega)} = d\left[\ln(R_{22}^2) + \ln\left(\frac{|A_{echo\,1}(\omega)|}{|A_{echo\,2}(\omega)|}\right)\right]^{-1} \tag{3}$$

In order to measure this value from the response matrix, the following process is followed. The first step is to perform a plane wave beamforming at zero incidence, meaning that the matrix K is summed along the emitters and the receivers without applying any delay law. The second step, leading to the determination of the coherent signal $A_{coh}(t)$, is to average over disorder, represented by the P positions of the array.

$$A_{coh}(t) = \langle \sum_{e,r} k_{er}(t) \rangle_P \tag{4}$$

Coherent waves calculated for both plates and both transducers are shown on Figure 2. It shows that according to grain size and frequency, the numbers of echoes that we are able to observe may vary. In the case of the SG sample, more than two echoes are detected whatever the frequency. On the contrary in the CG sample, the first two echoes are visible using the SONO12 array (c), but with the SONO65, the second one do not comes out of the noise (d). That means that in this plate, multiple scattering dominates very quickly at 3MHz.

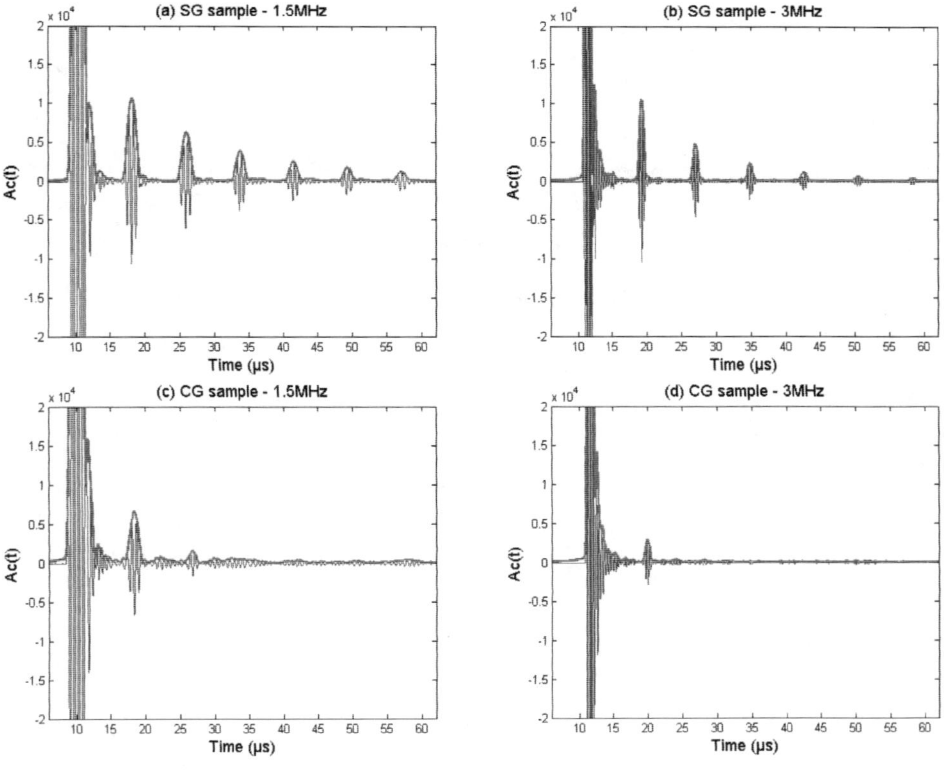

Figure 2. Coherent waves obtained for each couple sample-array: SG plate with SONO12 (a) and SONO65 (b); CG plate with SONO12 (c) and SONO65 (d).

The third step is to take the Fourier transform of the echoes. At last, using equation (3), we can calculate l_e, whose behaviour with frequency is shown on figure 3. For the SG plate, we observe the decrease of l_e with frequency. This is consistent with the fact that at lower λ/γ ratio, scattering effects are more important. For the CG plate, the same decrease is observed with the SONO12 array. Unfortunately, the thickness of the sample prevented the detection of the second back-surface echo, making it impossible to calculate l_e. This problem could be solved by testing samples with a lower thickness. Doing so will increase the ratio l_e/d and extend the numbers of visible echoes. However, this is not the best solution in the idea of a unique experimental setup as we will see in the following paragraphs how the back-surface echoes can limit the measure of the other parameters.

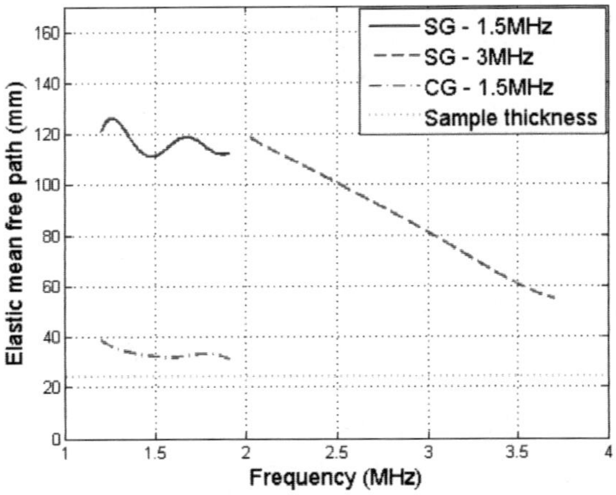

Figure 3. Elastic mean free path as a function of frequency in the steel samples.

3.2. Incoherent field and noise correlation distance

The recorded noise appears as a speckle pattern. By studying its spatial variations, it is possible to extract a characteristic size of its spots. We call this size the correlation distance d_c. It represents the fact that even if the noise appears to be random, two signals close to each other will show similarities. For simulation purpose, knowing its value will give information about the spatial behaviour of the phase term of the backscattered field.

During the study of the elastic mean free path, the calculation was made on the coherent wave. Here, we will be processing the incoherent wave. The aim is to get rid of the back-surface echoes, because their presence could compromise the calculation by introducing correlations at long distance that are not relevant in the speckle statistics. However, back-wall echoes are not the only contributions that are subtracted from the signals as it is known that the ballistic wave, which seems to propagate as if it was in a homogeneous medium, is only a part of the coherent wave [9].

The first step of the post-process is the same as for the mean free path, i.e. zero incidence beamforming is performed. But this time the sum is only made on emission, to obtain a B-scan where each signal is the sum of the coherent and the incoherent waves. To isolate the incoherent part $A_{inc}(t,r)$, the average of the B-scans over the positions of the array is subtracted:

$$A_{inc}(t,r) = \sum_e k_{er}(t) - \langle \sum_e k_{er}(t) \rangle_P \quad (5)$$

For every couple of elements r_i and r_j distant of δ from each other, we then calculate the cross-correlation at zero lag:

$$\varepsilon(\delta) = R_{x_i x_j}(0) = \int x_i^*(\tau) x_j(\tau)\, d\tau \quad (6)$$

where x_i and x_j are respectively the signals recorded by r_i and r_j. Finally, for each value of δ, ε is normalised by the number of occurrences of the inter-element distance. The resulting function is plotted on figure 4, for the CG plate and the SONO65 array (3MHz). It shows that the correlation of the signals along the length is well described by an exponential decay. By fitting the resulting curve with an exponential test function f, it is possible to deduce the value of d_c:

$$f(\delta) = a\, e^{-b.\delta} \quad , \quad (a,b) > 0 \quad \rightarrow \quad d_c = \frac{1}{b} \tag{7}$$

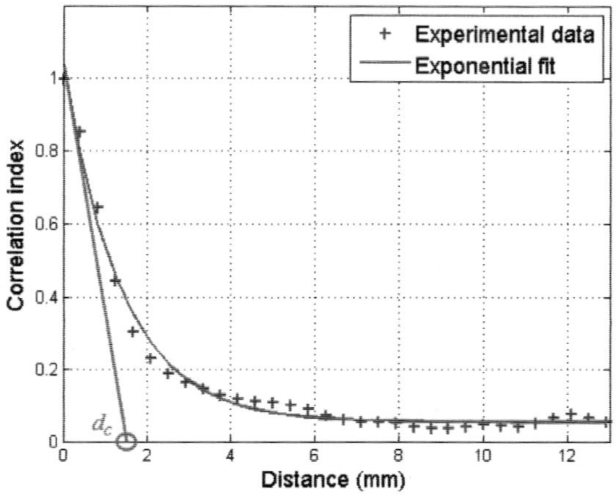

Figure 4. Normalised cross-correlation at zero lag as a function of inter-element distance for the CG sample at 3MHz.

The value of the correlation distance obtained by this method corresponds to the whole signal recorded by the array. However, its behavior with time is not determined that way. To access this information, the same calculation is done on 5-µs-long overlapping time windows δt:

$$\varepsilon(\delta, T) = \int_{T}^{T+\delta t} x_i^*(\hat{\tau}) x_j(\hat{\tau})\, d\hat{\tau} \tag{8}$$

The dependence of the correlation distance with time is calculated by applying the exponential fit at each step T of time. The result is shown on figure 5. In the CG sample (figure 5(a) and 5(b)) the spatial correlation of the structural noise is decreasing over time. However, it is not possible to ensure that it tends to zero, as the spatial resolution is limited by the pitch of the array. The local maxima that appear around 18µs and 25µs denote the persistence of the back-surface echoes even after the subtraction of the mean of the field. This is due to the deformation of the wavefront during the propagation in the scattering medium. The back-surface echo lost a part of its coherence and cannot be entirely eliminated. More averaging, or a more appropriate time-gating, would be required to gain a better correction.

In the SG sample (figure 5(c) and 5(d)) the same general behavior is obtained, but the prominence of the back-wall echoes over the incoherent wave is greater than in the CG samples. The long-distance correlations induced this way make the evolution of d_c poorly visible in these samples, and at these frequencies.

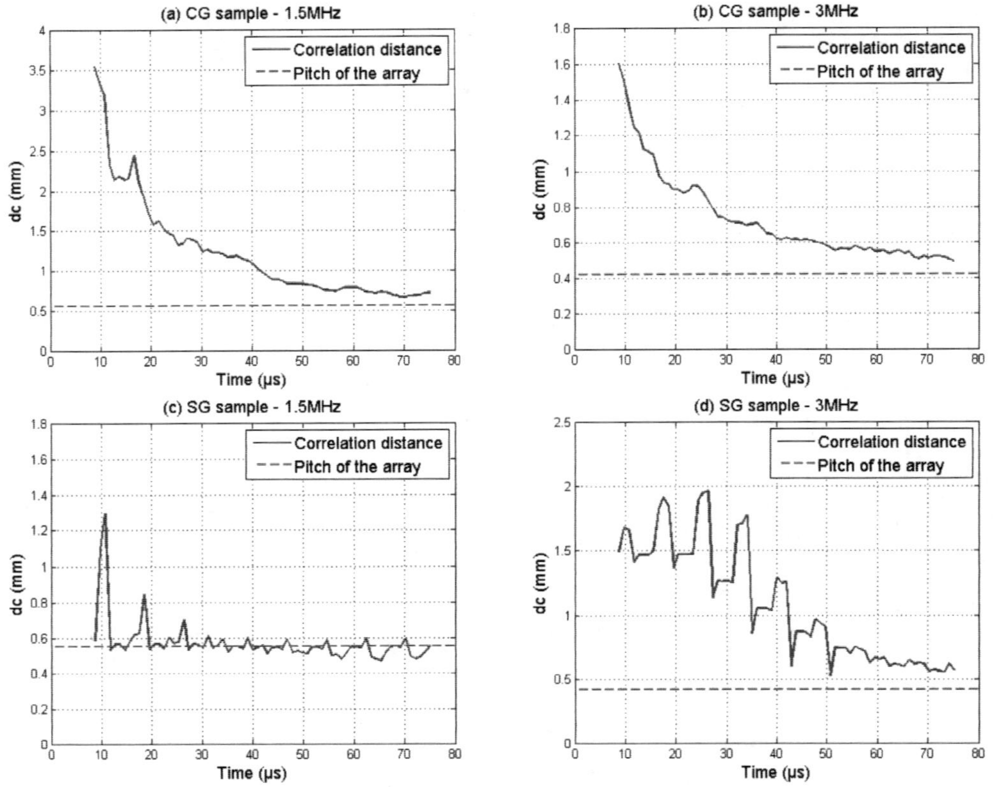

Figure 5. Correlation distance of the noise as a function of time. CG plate with SONO12 (a) and SONO65 (b); SG plate with SONO12 (c) and SONO65 (d).

3.3. Incoherent intensity and diffusion constant

As we saw in the first paragraph, after several mean free paths, the multiple scattering regime become dominant and the coherent wave is greatly attenuated. Indeed, the ultrasonic energy has been transferred to the incoherent wave, creating a halo spreading into the medium. The propagation of this halo can be treated as a random walk [4]. Under this approximation, called the diffusion approximation, the energy density over disorder obeys to the diffusion equation:

$$D\,\Delta\langle I(r,r',t,t')\rangle - \frac{\partial\langle I(r,r',t,t')\rangle}{\partial t} = \delta(r-r')\delta(t-t') \tag{9}$$

where *I* is the energy density of the wave, *D* is the diffusion constant and (*r'*,*t'*) is the position of the diffuse source in space and time. The diffusion constant is a dynamic parameter related to the growth speed of the diffusive halo. To infer the value of *D* from the response matrix, we will study the coherent backscattering effect. It highlights the long-time persistence of coherent interferences even in a disordered media. This effect corresponds to an enhancement in the intensity of waves that are backscattered along the incidence direction. Considering two scattering paths *p* and *q* into the medium (see figure 6(a)), the calculation of the average backscattered intensity leads to:

$$\langle I\rangle = \sum_p \langle |A_p|^2\rangle + \sum_p \sum_{q\neq p} \langle A_p^* A_q\rangle = I_{inc} + I_{coh} \tag{10}$$

where *A* is the amplitude of the considered path. The first term corresponds to the sum of individual intensities of each path. The contribution of this term to the total intensity will not present angular variations, at least in the far field and at small angles. The second term originates from the interference between different scattering paths. As an average over disorder is performed, the mean value of I_{coh} is zero in the general case because each path is uncorrelated with every other one. However, when source and receiver coincide, each multiple scattering path has a twin brother (its reciprocal counterpart, meaning that the same scatterers come into play but in the opposite order) and both path are necessarily in phase (figure 6(b)). Therefore at exact backscattering (i.e. $\theta = 0$, source = receiver) happens a coherent interference. Thus, $I_{coh} \approx I_{inc}$ and the mean intensity is higher by a factor 2. The angular width $\Delta\theta$ of the peak essentially depends on the conditions (far- or near-field), the wavelength and the transport parameters (D, l_e, l^*) of the multiple scattering medium.

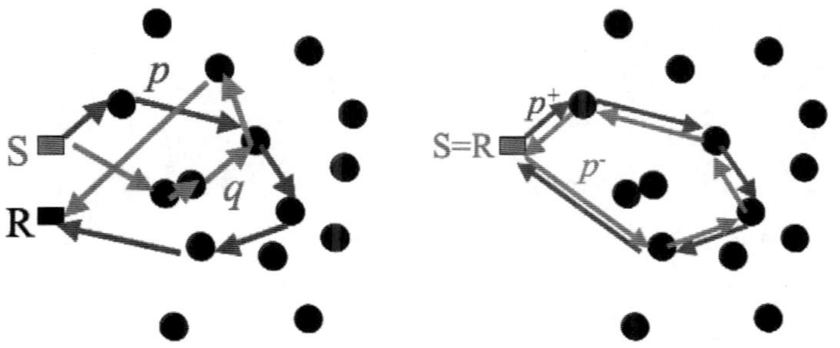

Figure 6. (courtesy of A. Aubry) Interferences (a) between uncorrelated scattering paths and (b) between reciprocal paths.

It has been shown [10] that under the far-field condition and in diffusive regime, the coherent backscattering spot width narrows with time:

$$\Delta\theta(t) \propto \frac{h}{k\sqrt{Dt}} \quad \text{with} \quad h \gg \sqrt{Dt} \tag{11}$$

where $\Delta\theta$ is the half-width at half-maximum of the coherent peak, h is the distance between the array and the sample, and k is the wavenumber in water. The recording of the backscattering peak and the study of its angular evolution with time will then give us access to D performing a linear fit of the inverse square of the half-width at half-maximum with respect to time:

$$\Delta\theta^{-2} = \frac{k^2 D}{\Gamma} t \tag{12}$$

where Γ is a factor which depends on the distance h and on the directivity of the transducers.

In this paragraph, we will compare two ways of computing the average backscattered intensity from the response matrix $K(t)$. These post-processing techniques are adapted from the work of A. Aubry on randomly located steel rods immersed in water [11]. The linear fit will be performed on both coherent backscattering peaks to compare the two methods.

The first and usual way to obtain the coherent peak is to work directly on the recorded signals $k_{er}(t)$. First, the signals are time shifted in reception to compensate for the different arrival times between the elements of the array. The intensity is then calculated by integrating the square of the signals on overlapping time windows. An average is then performed over all element couples in emission/reception that correspond to the same backscattering angle. Then another averaging is performed over the P positions of the array:

$$I_1(\theta, T) = \langle |k_{er}(t - \Delta t_{er})|^2_{t-\Delta t_{er} \in [T;T+\delta t]} \rangle_P \tag{13}$$

where Δt_{er} is the arrival time correction between the two signals, δt is the length of the time window, and $\theta = \text{Arctan}[(e\text{-}r).p/h]$ with p the pitch of the array. Figure 7(a) shows a result for the averaged backscattered intensity from the CG sample at 3MHz.

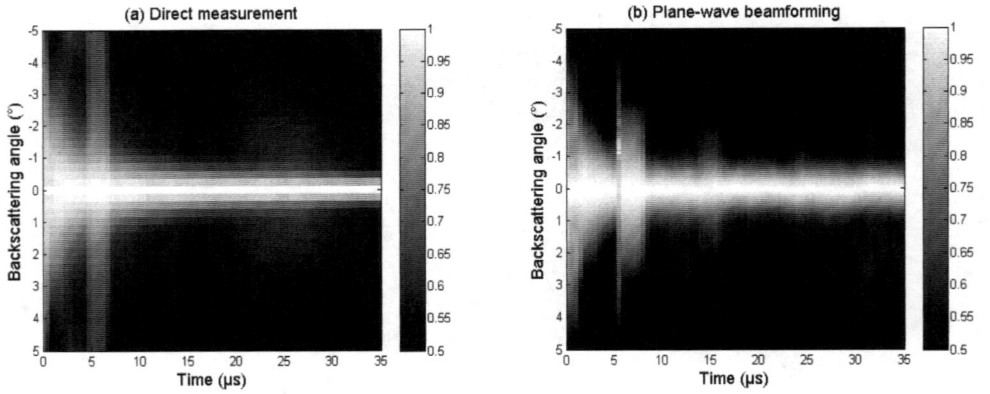

Figure 7. Dynamic backscattered intensity obtained from direct measurement (a) and from plane-wave beamforming technique (b) in the CG sample at 3MHz. At each time step, the intensity is normalised by its maximum along the backscattering angle.

In this figure, the intensity is normalized at each time step by its maximum for this time. This maximum always occurs around the 0° backscattering angle because this is where the coherent backscattering effect is most significant. The measurement is affected by a presence of a backwall echo around 6µs. For the rest of the time interval the width of the peak decreases regularly, as expected.

The other way to compute the coherent backscattering peak is to use plane-wave beamforming. As explained in section 2, the principle is to apply delay laws to the signals to simulate the emission and reception along different angles. Additionally, the use of plane-waves is a way to ensure that the far-field condition, required to observe the decrease of the peak's width, is fulfilled. It is possible to work in the time domain or in the Fourier space. In the time domain, a small variation of angle implies the application of a delay that can be lower than the sampling time. On the contrary, applying delay laws in the frequency domain reduces the operation to a simple matrix multiplication. For this reason, we chose to work in Fourier space. First, a fast Fourier transform gives the response matrix $K(f)$ in the frequency domain. Then, considering the emission along the angle α and the reception along the angle β, the delay laws are calculated for each element of the array with the first transducer taken as reference:

$$\begin{cases} \tau_\alpha(e) = \dfrac{(e-1)p\sin\alpha}{c} \\ \tau_\beta(r) = \dfrac{(r-1)p\sin\beta}{c} \end{cases} \tag{14}$$

where c is the velocity of the wave in the coupling medium. Next, we take the Fourier transform of these laws:

$$\begin{cases} \tilde{\tau}_\alpha(f,e) = e^{-2i\pi f \tau_\alpha(e)} \\ \tilde{\tau}_\beta(f,r) = e^{-2i\pi f \tau_\beta(r)} \end{cases} \tag{15}$$

After that, we calculate the matrix multiplication for each frequency to obtain the signal integrated by the array after the emission along α and the reception along β.

$$s_{\alpha,\beta}(t) = TF^{-1}\left[\tilde{\tau}_\alpha(f,e) * \tilde{k}_{er}(f) * {}^t\tilde{\tau}_\beta(f,r)\right] \tag{16}$$

The dynamic intensity $I_2(\theta,T)$ is finally obtained by integrating the square norm of $s_{\alpha,\alpha+\theta}(t)$ over overlapping time windows and by averaging over the emission angle α and the P positions of the array. The result calculated in the CG sample at 3MHz is shown on figure 7(b).

$$I_2(\theta,T) = \langle |s_{\alpha,\alpha+\theta}(t)|^2_{t\,\epsilon\,[T;T+\delta t]} \rangle_{\alpha,P} \tag{17}$$

The study of the direct measurement peak (figure 7(a)) shows the enhancement by a factor 2 of the backscattered intensity. Moreover, with this method, the angular resolution is limited by the pitch of the array. Indeed, it is impossible to get information below the angle p/h, limiting the range of calculation of D.

Using plane-wave beamforming, we see that there is less energy along the higher backscattering angles, meaning that the signal-to-noise ratio has been increased. It is the consequence of the use of every impulse responses contained in the response matrix to determine a single signal. Additionally, it is theoretically possible to improve the angular resolution and to access long time information about the angular variation of the backscattering peak. However, a saturation phenomenon is quickly observed. Indeed, in the far field setup, the angular resolution is dependant of the aperture of the array and the saturation occurs when the width of the peak reach λ/Np. In our experimental setup, the size of the available array (≈27mm), associated with a 3MHz pulse, leads to a limit of 1°. This value is consistent with the experimental result presented on figure 7(b) as we see that the total width at half-maximum (i.e. for an amplitude of 0.75) is 1°. Increasing the size of the array should improve the angular resolution.

Finally, we see that with both techniques the presence of the first back-surface echo is a limiting factor. On the one hand, its presence prevents the visualisation of the peak evolution. On the other hand, the behaviour of noise could be modified due to the backwall reflection that will bring more energy to the diffuse field as it propagates again through the medium. Thus, the determination of D will be done in the time interval before the apparition of the back-surface echo.

Before extracting the value of D from the coherent backscattering peak, we have to verify the applicability of equation (11). Indeed, this equation is only valid in the diffusion approximation. To that aim, we verify the time dependence of the peak's width decrease. If the diffusion approximation is valid, the slope of the linear fit of $\log(\Delta\theta)$ with respect to $\log(t)$ must be equal to -0.5. The result of the calculation is plotted on figure 8.

For direct measurement, the slope is inferior to the expected value, but with beamforming, we obtain a good agreement with the theory. That means that the determination of the diffusion constant is more likely to be precise with the beamforming treatment. It also indicates that though there is undoubtedly multiple scattering, the truly diffuse regime is probably not attained. In that case the energy spreads from the source in a manner not as simple as $\sqrt{Dt}$ and is more accurately described by the radiative transfer equation [3].

Finally, D is obtained by the linear regression of the inverse square of the half-width at half-maximum with respect to time (equation (12)). The slope of this regression leads to the diffusion constant as:

$$D = \frac{ah^2\Gamma^2}{k^2} \tag{18}$$

The results of the two regressions are plotted on figure 8. The values obtained are:

$$\begin{cases} D_1 = 7.5\,mm^2/\mu s \\ D_2 = 6.3\,mm^2/\mu s \end{cases} \tag{19}$$

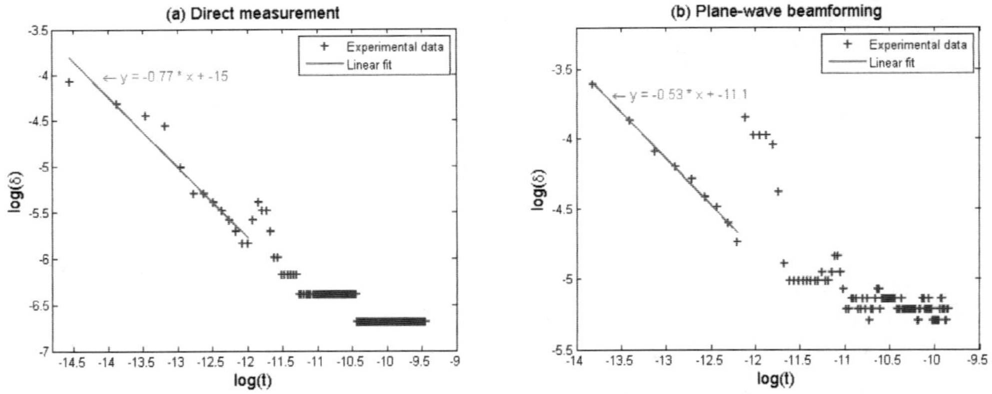

Figure 8. Verification of the diffusion approximation validity to determine the diffusion constant: (a) direct measurement; (b) plane-wave beamforming. The calculation of D is performed before the first back-surface echo. The two limitation phenomena occurring at long time are observed.

The same order of magnitude is obtained, which seems to show that the two methods are consistent. However, due to the higher signal-to-noise ratio and the better validation of the diffusion approximation condition, D_2 is probably more trustworthy. Yet we insist on the fact that these values are to be taken as orders of magnitude rather than precise measurements, given the limited range of time lags and the questionability of the diffusion approximation in that context. Other experimental techniques to characterise the diffusion constant exist. Particularly, a study of the time-of-flight function recorded in transmission is often used [4,12]. The comparison of our values with the results of these techniques in stainless steel samples could be interesting to do.

4. Conclusion

In this study, various post-processing techniques have been applied to the response matrix of a scattering medium, leading to the calculation of three parameters that are being used as entry parameters in an ultrasonic noise simulation method currently in progress at the CEA LIST. We showed that it is possible to obtain the elastic mean free path l_e, the correlation distance of the noise d_c, and the diffusion constant of the medium D by performing only one experimental measurement with a linear phased array. The main advantage of using the response matrix is that all information about the array and the inspected medium is contained in the impulse responses. It is then very easy to test other post-processing techniques without the need of carrying out other experiments. Another advantage is that fewer experiments are carried out and then more time can be devoted to post-processes. The main disadvantage, compared to measurements which are optimised for the calculation of a target parameter, is that going through the response matrix implies larger stored data and larger computation times.

The determination of all three parameters highlighted the difficulties dealing with the back-surface echoes. The presence of these echoes is needed to determine the elastic mean free path, but appears to be the main limitation when trying to calculate the correlation distance of the noise and the diffusion constant. Indeed, l_e characterise the coherent wave (i.e. the signal that resists to averaging processes), contrary to d_c and D that are related to the incoherent wave, and thus characterising the random noise that is classically eliminated by ensemble averaging.

Future work will study the link between dc and the microstucture, as well as refine the estimated values for D in similar samples. This will require either a new set of experiments with deeper samples, or, in order to study the short-time evolution of $\Delta\theta$, maybe develop a theoretical approach based on the radiative transfer equation, which is evidently much more complicate to handle than the simple diffusion approximation.

Journal of Physics: Conference Series **457** (2013) 012003 doi:10.1088/1742-6596/457/1/012003

Finally, with the study of the coherent backscattering effect, we presented an alternative to the direct measurement of the backscattered ultrasonic intensity, by using plane-wave beamforming. Here again, the computation time is increased, but it comes with higher signal-to-noise ratio and greater angular resolution. Even if our arrays were not the best suited to emphasise this latter benefit due to the rapid saturation of the peak width, the use of larger ones can improve the quantitative determination of the diffusion constant and the study of the influence of back-surface echoes on the dynamic of the peak at longer times.

References

[1] Dorval V, Jenson F, Corneloup G and Moysan J 2010 *Review of Progress in QNDE* **29** 1309

[2] Feuilly N, Dupond O, Chassignole B et al., 2009 *Review of Progress in QNDE* **28** 1216

[3] Turner J A and Weaver R L 1994 *J. Acoust. Soc. Am.* **96** 3675

[4] Page J, Schriemer H P, Bailey A E and Weitz D A 1995 *Phys. Rev. E* **52** 3106

[5] Derode A and Fink M 1997 *J. Acoust. Soc. Am.* **101** 690

[6] Angelsen B 2000 *Ultrasound imaging: waves, signals and signals processing* (Trondheim: Emantec)

[7] Tourin A, Derode A, Peyre P and Fink M 2000 *J. Acoust. Soc. Am.* **108** 503

[8] Wu P and Stepinski 2000 *Ultrasonics* **38** 481

[9] Derode A, Tourin A and Fink M 2001 *Phys. Rev. E* **64** 036605

[10] Tourin A, Derode A, Roux P, van Tiggelen B A and Fink M 1997 *Phys. Rev. Lett.* **79** 3637

[11] Aubry A and Derode A 2007 *J. Acoust. Soc. Am.* **121** 70

[12] Anugonda P, Wiehn J S and Turner J A 2001 *Ultrasonics* **39** 429

Ultrasonic attenuation measurements at very high SNR: Correlation, information theory and performance.

Richard Challis, Vladimir Ivchenko and Raied Al-Lashi

Electrical Systems and Optics Division, Faculty of Engineering
University of Nottingham, Nottingham NG7 2RD

E-mail: richard.challis@nottingham.ac.uk

Abstract. This paper describes a system for ultrasonic wave attenuation measurements which is based on pseudo-random binary codes as transmission signals combined with on-the-fly correlation for received signal detection. The apparatus can receive signals in the nanovolt range against a noise background in the order of hundreds of microvolts and an analogue to digital convertor (ADC) bit-step also in the order of hundreds of microvolts. Very high signal to noise ratios (SNRs) are achieved without recourse to coherent averaging with its associated requirement for high sampling times. The system works by a process of *dithering* – in which very low amplitude received signals enter the dynamic range of the ADC by 'riding' on electronic noise at the system input. The amplitude of this 'useful noise' has to be chosen with care for an optimised design. The process of optimisation is explained on the basis of classical information theory and is achieved through a simple noise model. The performance of the system is examined for different transmitted code lengths and gain settings in the receiver chain. Experimental results are shown to verify the expected operation when the system is applied to a very highly attenuating material – an aerated slurry.

1. Introduction

This paper describes an ultrasonic spectrometer for measurements of compression wave attenuation and phase speed as functions of frequency at very much higher signal to noise ratios (SNRs) than would be expected from a conventional system based on a high voltage pulser-receiver. The motivation was a requirement to make attenuation measurements in aerated slurries contained in pipework with diameters up to 150mm. The expected ultrasonic propagation path loss was of the order of 400dB. We have shown [1] that the ideal path loss which maximizes SNR in attenuation data is 1Np, equivalent to 8.68dB, a factor of 46 below the stated 400dB. A further constraint was that the test slurries would be flowing along the pipe with the likelihood that the physical properties of the material between the interrogating transducers would not be stationary. This implies that insufficient time would be available to apply coherent averaging to the received signal in order to improve the SNR. In addition, the physical environment in which the test system might be located could be subject to fire or explosion hazards, and this set a further requirement that the electrical signals used to excite the transmitting transducer should have a peak amplitude in the order of 1V.

The solution we have adopted to this combination of adverse SNR conditions and measurement constraints is a spectrometer device in which a code in the form of a low voltage pseudo random binary sequence (PRBS) is used as the transducer excitation signal, and the received signal is extracted

Content from this work may be used under the terms of the Creative Commons Attribution 3.0 licence. Any further distribution of this work must maintain attribution to the author(s) and the title of the work, journal citation and DOI.
Published under licence by IOP Publishing Ltd

by a correlation operation. In the following sections we outline the basic operation of the system and demonstrate its performance. A number of details in relation to the design of the system are to be found in [2].

2. System operation

Codes and correlation have been used successfully for system identification for many years [3-11]. We have established that although many different types of code, or indeed analogue noise signals, could be used as transmission signals [12] in correlation spectrometers, that which gives the best performance is the Golay code [13]; this is a digital code and broad band in nature. Figure 1 shows the basic system arrangement.

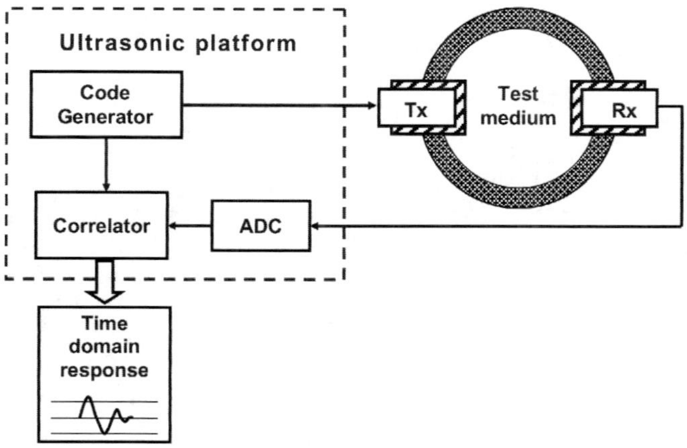

Figure1. The basic system arrangement; Tx and Rx are the transmitting and receiving transducers mounted on either side of the pipe containing the test medium.

The Golay code excites the transmitting transducer and a corresponding acoustic signal passes through the test medium and is received by a second transducer, the output of which is digitized (quantised) by an analogue to digital converter (ADC). It is then cross-correlated with a copy of the transmitted code to produce the impulse response of the whole signal path limited by the bandwidth of the transmitted code. In practice this operation is done twice with two Golay codes (A and B) which are complementary to each other in a binary sense. The auto-correlation functions (ACFs) of the two codes both feature a peak at zero lag as would be expected on the basis that they are both random. However, they are not ideally random due to their finite length, and this results in so-called self noise in their ACFs after the principal peaks at zero lag. The special feature of Golay codes is that the self noise in code A is in antiphase with that of code B. Thus if the ACFs of the two codes are added together the principal peaks add constructively whilst the two self noise components cancel each other. This feature of self noise cancellation usefully follows through to the cross-correlation process – the self noise components in the two cross-correlation functions (CCFs) cancelling in summation. The double transmit-receive-correlate process is illustrated in figure 2.

The resulting output is equivalent to the pulse signal which would have been received using a conventional pulser-receiver system, except that the SNR is far superior. If the raw SNR of either of the received signals (A or B) is SNR_1, then the SNR after the correlation and addition process [2] is

$$SNR_2 = SNR_1\sqrt{2NR} \qquad (1),$$

where N is the length of the Golay code expressed as the number of binary state changes, and

$$R = \frac{f_s}{f_{ex}} \quad (2),$$

where f_s is the ADC sampling frequency and f_{ex} is the average frequency of clock state changes associated with the code.

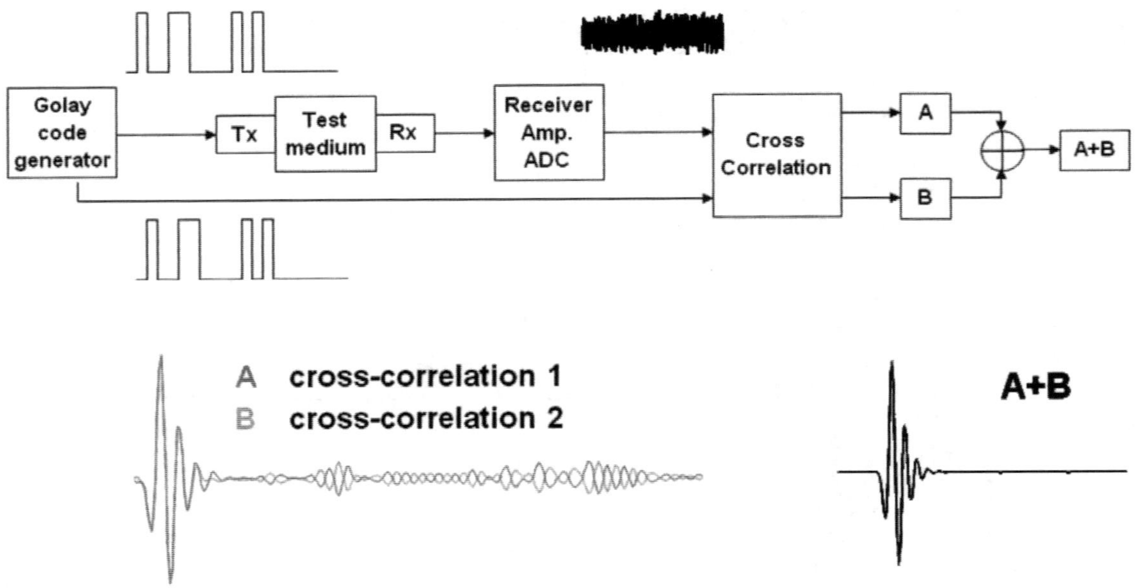

Figure 2. (Top) The double transmit-receive-correlate process. (Bottom) The two cross-correlograms, A and B, showing (left) their self noise components after the initial peaks at zero lag, and (right) the sum of the two with the self noise components cancelled.

3. Initial assessment of performance

The system was implemented in hardware based on a field programmable gate array (FPGA) and applied to measurement of the highly attenuating slurries. The SNR performance was excellent – the system was able to capture a received signal with an amplitude of 70nV against a noise level of 335μV rms and an ADC bit-step equivalent to 488μV, see figure 3. We now describe the implicit signal manipulations that yield this performance and the implications of these on system design.

4. Implicit signal manipulations

The first consideration is how such a small signal can be digitised successfully with a large bit-step in relation to its amplitude. The central phenomenon is that the signal at the input of the ADC adds to the system noise and effectively rides on this noise as it is digitised. This enables signal samples to enter the dynamic range of the ADC, a process known as *dithering* [14-16]. The integration process implicit in the cross-correlation operation (accumulation in digital terms) reduces this noise considerably, leaving the required signal component expressed over several bits of the ADC output, even though its initial amplitude was very much less than a single ADC bit-step.

A central issue which impinges on the design of this type of system is the optimum amplitude for the system noise at the ADC input – the noise on which the input signal rides. Here, if the noise were too great then the processed signal would be contaminated by that noise. Conversely, if the noise were too small the dithering process would be compromised, with distortion and excessive noise at the ADC output. This problem has been addressed in a number of early works in information theory [17- 20], the principal results of which we will now summarise.

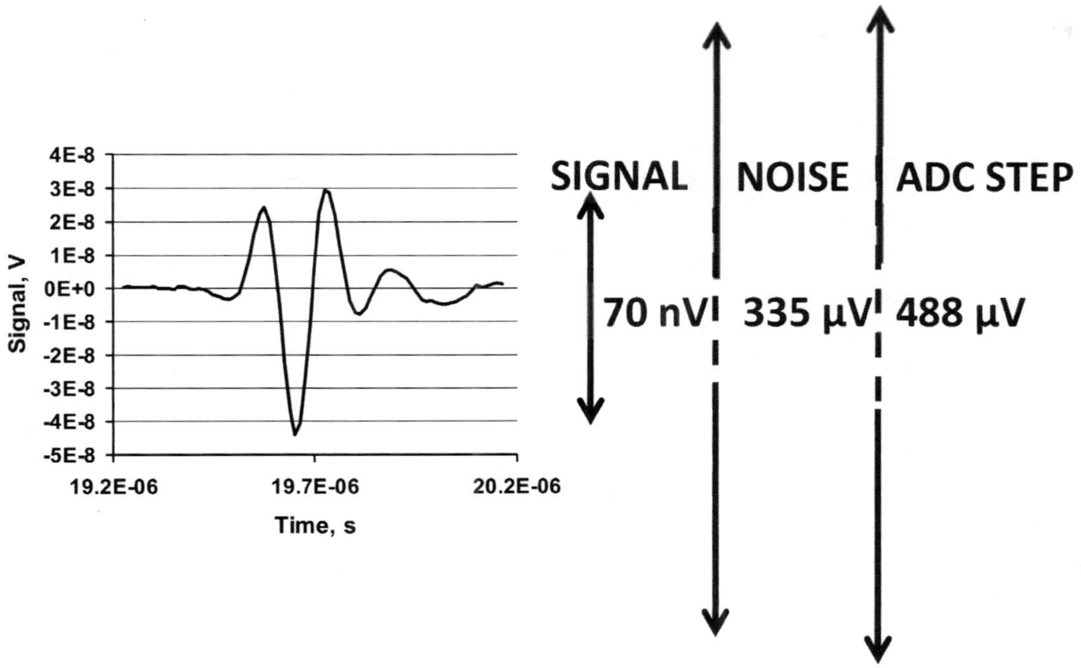

Figure 3. Illustration of a received signal of very low amplitude compared to the noise level at the ADC input and the ADC bit-step.

5. Optimum noise level

Figure 4 illustrates the quantisation process that underpins the operation of the ADC; the horizontal axis represents the input signal amplitude and the vertical axis represents the digitised result. The step amplitude corresponding to one ADC bit is Δ. Figure 5 shows a typical probability density function (PDF) for Gaussian noise with vertical arrows representing the ADC quantisation steps. Expressing the quantisation process in this way leads to the idea that it can be considered as a sampling operation in the amplitude domain. If the noise standard deviation σ were small then the digitised version would be under sampled, and vice-versa. The relationship between the ADC bit resolution and the noise amplitude can be usefully considered by taking the Fourier transform of the amplitude samples on Fig. 5. Here, the input domain to the transform is *amplitude*, and the transformed result is a domain in *reciprocal amplitude*. The Fourier transform of the PDF $p(x)$ is its characteristic function (CF) given by

$$\Phi(u) = \int_{-\infty}^{\infty} p(x)e^{iux}\,dx \tag{3}$$

There is a fundamental quantising theorem [18, 19] which states that if $\Phi(u)$ is band limited in the reciprocal amplitude (u)-domain such that

$$\Phi(u) = 0 \quad \text{for } |u| \;\rangle\; \frac{\pi}{\Delta} = \frac{\Psi}{2} \qquad (4),$$

where Ψ is the sampling rate in the amplitude domain, then the CF and PDF of x_A can be derived from the corresponding digitized data. This statement is equivalent to the Nyquist theorem for time to frequency transforms.

Figure 6 shows the transformed result (eq. (3)) for Gaussian noise with three different noise amplitudes σ, each normalised to the quantisation step Δ. On figure 6 the ideal sampling condition in the amplitude domain is when the quantisation step is equal to the noise standard deviation,

$$\frac{\sigma}{\Delta} = 1 \qquad (5)$$

The curve for $\sigma/_{\Delta} = 0.5$ is under sampled and that for $\sigma/_{\Delta} = 10$ is over sampled. We have simulated this digitising process [2] in order to assess the extent to which the input noise passes through the ADC to produced digitised *output* noise of the same amplitude for different Golay sequence lengths N. Figure 7 shows the result.

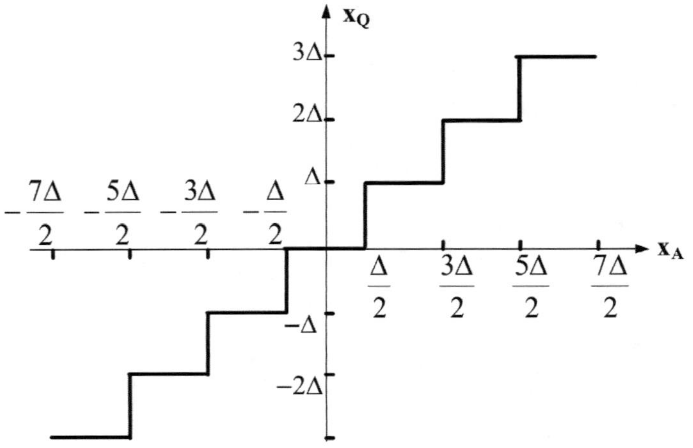

Figure 4. The quantisation process. The horizontal axis represents the input signal amplitude and the vertical axis represents the digitised result.

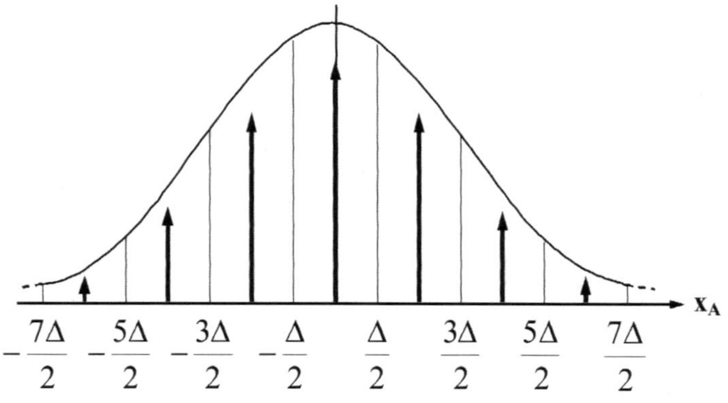

Figure 5. PDF of Gaussian noise showing amplitude samples at intervals of the ADC bit step Δ.

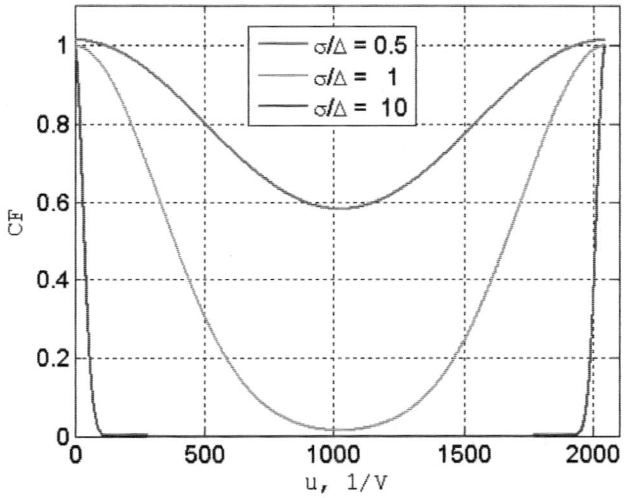

Figure 6. CF of the quantised noise on figure 5 in the inverse-amplitude (u) domain, showing the result for three quantisation rates.

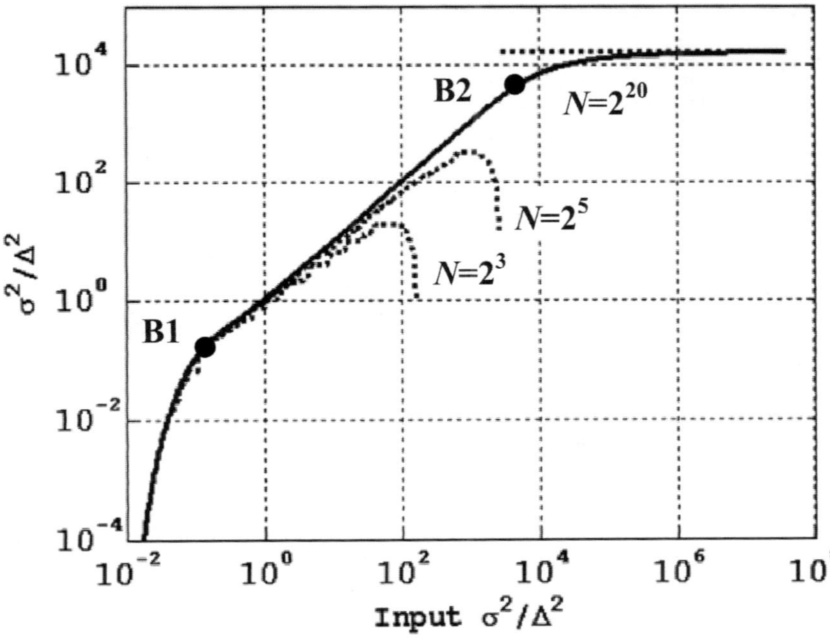

Figure 7. Relationship between the analogue input noise amplitude (horizontal axis) and the digitised noise amplitude (vertical axis), both normalised to the ADC bit step Δ. The results for three sequence lengths N are shown.

The principal curve between break points B1 and B2 matches a result derived many years ago by Brown [17]. At the low amplitude end B1 is the point below which the input noise amplitude is so low that the probability of any sample of it entering the ADC range is negligible. At the higher amplitude, break point B2 corresponds to a significant probability that samples of the input noise will exceed the dynamic range of the ADC. It is also clear from the figure that to achieve the 'ideal' characteristic between B1 and B2 high sequence lengths N are required. For short sequences, successful transmission through the ADC is possible only for a limited range of noise amplitudes – with σ/Δ around 6 for $N = 2^3$, and around 30 for $N = 2^5$. We established earlier that the optimum noise level was around $\sigma/\Delta = 1$, which when referred to figure 7 would provide for successful operation with short sequences. We would expect that results related to these would arise when a deterministic (information) signal passed through the system, added to the noise in the dithering process.

6. Signal transmission

The same simulation tool as was used to analyse noise transmission was employed to simulate information signal transmission through the system for various levels of noise at the input to the ADC, and for different Golay sequence lengths. The results on figure 8 show that, for input noise amplitudes below around 0.5Δ, signal transmission falls dramatically; this is because the noise amplitude is not sufficient to project enough samples on the input signal into the dynamic range of the ADC. For noise levels above this the signal transmission is unity up until a limiting noise amplitude which increases with Golay sequence length. At this limit the implicit integration in the correlation process is insufficient to reduce the noise to the required insignificant level; only an increase in sequence length can overcome this, with the disadvantage that longer data acquisition times would result.

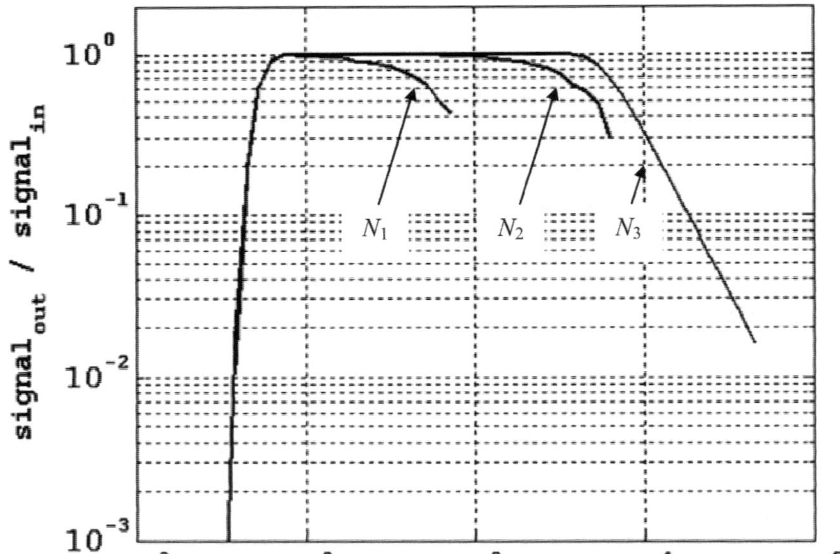

Figure 8. Information signal transmission through the system: The ratio of the amplitudes of the digitised signal and the input analogue signal versus normalised noise amplitude σ/Δ for various sequence lengths $N_1 = 2^{12}$, $N_2 = 2^{18}$, and $N_3 = 2^{30}$.

7. Choice of input noise amplitude

We have seen that the input noise together with the sequence length determine the transmission of both noise and signal through the ADC. If the noise amplitude is too low both signal and noise will be degraded; excessively high noise amplitude has two adverse effects: (i) The dynamic range limit of the ADC is exceeded with consequent distortion, and (ii) Longer sequence lengths will be required so as to improve the SNR in the correlation. In order to check our earlier decision on an optimum noise level we have simulated the SNR performance of the system and figure 9 shows the result. We see the highest SNR is achieved for $\sigma/\Delta = 0.3$, and that lower noise levels bring about a rapid fall in SNR. The SNR also decreases at higher noise levels, but at a lower rate. In relation to system design it therefore makes sense to err on the side of safety and to set the noise level just above 0.3 in order to protect against an unforeseen reduction in noise level in the physical electronic hardware. We have therefore maintained the noise level such that its standard deviation is equal to one ADC bit-step, $\sigma = \Delta$. The source of this required noise will be discussed in a later section.

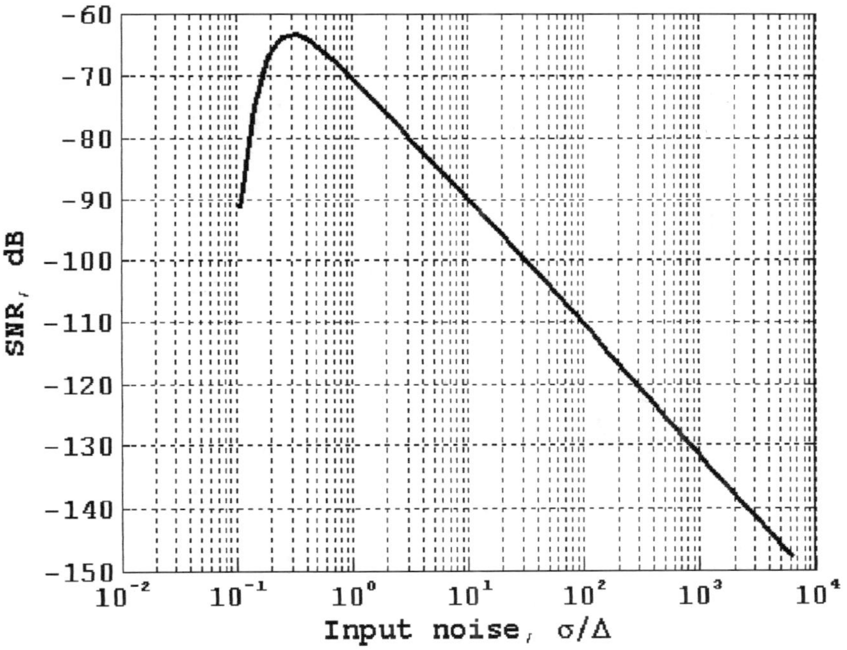

Figure 9. SNR for an input signal of 5 μV amplitude versus input noise normalised to the quantisation step for a sequence length $N=2^{30}$.

8. Minimum input signal

The performance observed for very low signal levels was illustrated in figure 3. It is to be expected that, given an appropriate level of dithering noise, the minimum signal level that is receivable will reduce as the sequence length is increased. We have used the simulation to establish the form of this relationship, and figure 10 shows that this is linear; the cross on the figure indicates the operating conditions of figure 3. It is also clear from the figure that even smaller signals would be receivable with increased lengths – indeed, if the sequence length tended to infinity then the minimum receivable signal would tend to zero. However, high sequence lengths imply long acquisition times, so there is a practical limit which will derive from a combination of the time available for measurement and the stationarity of the physical properties of the test medium.

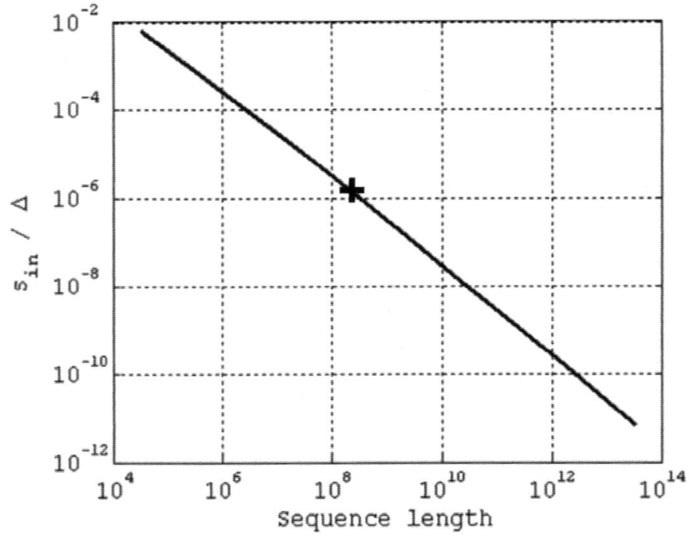

Figure 10. Minimum receivable signal level versus sequence length *N*. The cross indicates the operating conditions of figure 3.

9. Practical considerations in relation to input noise

The ADC bit-step in the current system is 488μV and so for $\sigma = \Delta$ we would require noise at the ADC of this amplitude. In most practical radio frequency electronic systems the commonly used interconnection impedance is 50Ω□□, and with an operating bandwidth of 40MHz this would result in a Johnson noise amplitude of 5.75μV, clearly not enough. Other extraneous noise sources would be expected to exist, and in the current system we have measured the actual noise level at 14.6μV, still very much below the required value. Whilst input amplifiers are normally used to improve SNR, we have used an amplifier/filter combination as a source of dithering noise, figure 11.

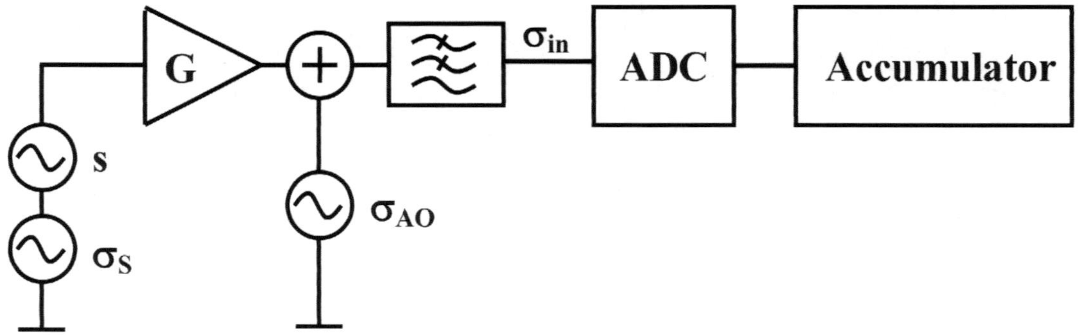

Figure 11. Overall system model. **G** is the gain block (amplifier) which is followed by the filter (insertion loss 10.4dB), the ADC, and the accumulator which represents the integration associated with the cross correlation process. The signal source is **s** and σ_s is the noise at the input; σ_{AO} is the intrinsic noise of the amplifier and σ_{in} represents the noise at the input to the ADC.

The filter insertion loss was 10.4dB, corresponding to factor 0.3. To achieve $\sigma = 488\mu V$ we would thus require 1627μV of noise at the filter input. In the arrangement of figure 11 the intrinsic noise of the amplifier was measured as $\sigma_{AO} = 335\mu V$ over the 40MHz bandwidth. The source noise at the input of the amplifier was $\sigma_s = 14.6\mu V$. The noise at the filter input is

$$\sigma_{\inf}^2 = G^2\sigma_s^2 + \sigma_{AO}^2 \qquad (5)$$

Re-arranging this we get the amplifier gain as

$$G^2 = \frac{\sigma_{\inf}^2 - \sigma_{AO}^2}{\sigma_s^2} \qquad (6)$$

Setting $\sigma_{inf} = 1627\mu V$, $\sigma_{AO} = 335\mu V$ and $\sigma_s = 14.6\mu V$ yields $G = 40.7dB$. The amplifier will also magnify the signal at its input, s, giving further advantage in terms of the overall SNR, but only up to a gain limit of

$$G = \frac{\sigma_{AO}}{\sigma_s} \qquad (7)$$

Beyond this point the SNR at the output will be dominated by the signal and noise amplitudes at the input of the amplifier. If an amplifier gain setting lower than the required 40.7dB were to be used, the noise input to the ADC would be inadequate for successful dithering, resulting in diminished overall SNR, as established in figure 9.

10. Overall system SNR

We required to establish the noise performance of the system for various combinations of transmission voltage, V_{PRBS}, sequence length N and factor R in eq. (2); we defined a 'quality factor' thus

$$Q_1 = 10\log(2NRV_{PRBS}^2) \qquad (8)$$

In order to compare the performance with that of a conventional pulser-receiver system we defined a second quality factor as

$$Q_2 = 10\log(V_P^2 M) \qquad (9),$$

where V_P is the transmission pulse amplitude and M is the number of coherent averages applied to the received signal before the SNR calculation. The maximum values for the pulser output and M were 100V and 1000 respectively, representing typical operating conditions and giving a maximum value of $Q_2 = 70$ dB. For the PRBS system the settings were $V_{PRBS} = 2V$, $R = 2$, and a maximum value for $N = 2^{33}$. The input signal level was 260nV and measurements were done for various amplifier gains in the range 5dB to 43.5dB, the maximum attainable for the device used. The results of these measurements are shown on figure 12 from which we note that the SNRs achievable with the coded system set to a transmitter voltage of 2V are far superior to those of the pulsed system with a transmitter voltage of 100V. The figure also shows the effect of the amplifier gain, the SNR increasing as gain is increased and maximising when G was around 40dB, as expected from eq. (6).

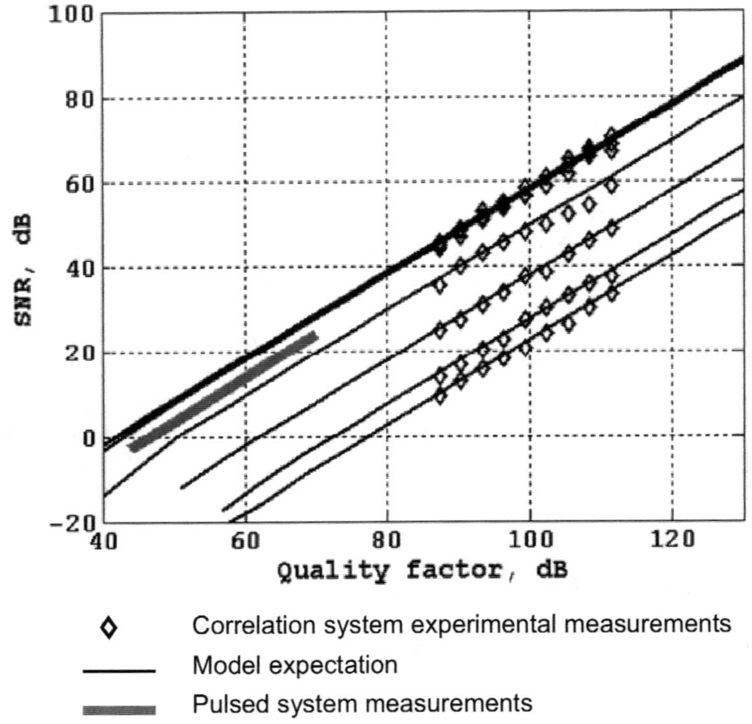

Figure 12. Overall system SNR plotted versus factor $QI = 10\log_{10} V_{PRBS}^{2} \times 2NR$ for amplifier gains of (bottom to top) 0, 5, 15, 25, 36, 40 and 43.5 dB. The lines are expectation from the system model and the symbols are experimental measurements. The lines for the three highest gains are approximately coincident. The emboldened solid line is the result for the conventional pulser-receiver system.

11. Concluding remarks

We have discussed the theory and performance of a coded sequence ultrasonic spectrometer and have demonstrated that it can receive signals which are very much lower in amplitude than both the noise at the input of the system and the ADC bit-step. This performance was achieved by a process of dithering whereby the input signal rides on the input noise to reach into the dynamic range of the ADC. The setting of the amplitude of this noise was crucial to the effective operation of the system and necessitated an input amplifier to supply noise of the appropriate amplitude – the standard deviation being equal to the ADC bit-step. In relation to system design, it is necessary to consider all of the noise sources at the ADC input – Johnson noise associated with the input impedance, extraneous input noise, and the intrinsic noise associated with the amplifier.

It should be noted that the dithering concept could also be applied to pulser-receiver systems, in combination with coherent averaging of the received signal to improve SNR. Whilst such a strategy may seem easier than the complexities of coded systems, it has the disadvantage that very long data capture times would be required on account of the necessary delays between successive acquisitions to allow time for acoustic reverberations between the transducers to decay to an acceptable level.

Finally, this outline has not considered the details of the underpinning electronic circuit design, the reason being confidentiality arrangements associated with its commercial exploitation. However, we can disclose that the FPGA, the programmable gain amplifier, filter and ADC were all off-the-shelf components. The ADC clock runs at 80MHz and the code clock is synchronised to this and runs at 40MHz; The system can be programmed to operate more slowly by successive binary division of both clocks. The Golay code length can be programmed in the range 2^8 to 2^{33}. For the future we aim to miniaturise the system and to provide a multiple channel facility for studies of chemical substances that are both flowing and undergoing chemical reaction.

References

[1] Kalashnikov A N and Challis R E 2005 Errors and uncertainties in the measurement of ultrasonic wave attenuation and phase velocity. *IEEE Trans. UFFC* **52(10)** 1754-1768.

[2] Challis R E and Ivchenko V G 2011 Sub-threshold sampling in a correlation-based ultrasonic spectrometer. *Meas. Sci. Technol,* **22** 1-12.

[3] Bilgutay N M Furgason E S and Newhouse V L 1976 Evaluation of a random signal correlation system for ultrasonic flaw detection *IEEE Trans. Sonics and Ultrason* SU-23(5) 329-333.

[4] Carpentier M H 1970 Using random functions in radar applications *De Ingenieur* 46 (13) E 166.

[5] Cook C E 1960 Pulse Compression – key to more efficient radar transmission *Proc. IRE* 48 310-316.

[6] Craig S E Fishbein W and Rittenbach O E 1962 (April) Continuous-wave radar with high range resolution and unambiguous velocity determination *IRE Trans. Mil. Electronics* MIL-6 153-161.

[7] Ding Z X and Payne P A 1990 A new Golay code system for ultrasonic pulse echo measurements *Meas. Sci. Technol.* **1** 158-65.

[8] Furgason E S Newhouse V L Bilgutay N M and Cooper G R 1975 Application of random signal correlation techniques to ultrasonic flaw detection *Ultrasonics* **13** 11-17.

[9] Hayward G and Gorfu Y 1988 A digital hardware correlation system for fast ultrasonic data acquisition in peak power limited applications *IEEE Trans. Ultrasonics, Ferroelectrics and Freq. Control* **35(6)** 800-808.

[10] Lee B B and Furgason E S 1982 An evaluation of ultrasound NDE correlation flaw detection systems *IEEE Trans. Sonics and Ultrasonics* **SU-29(6)** 359-369.

[11] Lee B B and Furgason E S 1983 High-speed digital Golay code flaw detection system *Ultrasonics* **21(4)** 153-163.

[12] Phang A P Y, Challis R E, Ivchenko V G and Kalashnikov A N 2008 A field programmable gate array-based ultrasonic spectrometer. *Meas. Sci. Technol.* **19(4)**, 1-13.

[13] Golay M J E 1961 Complementary series *IRE Trans. Inform. Theory* **7** 82-87.

[14] Lipshitz S P Wannamaker R A and Vanderkooy J 1992 Quantization and dither: A theoretical survey *J. Audio Eng. Soc.* **40** 355-375.

[15] Carbone P and Petri D 1994 Effect of additive dither on the resolution of ideal quantizers *IEEE Trans. Instrum. and Meas*. **43(3)** 389-396.

[16] Gray R M and Stockham T G 1993 Dithered quantizers *IEEE Trans. Inform. Theory* **39(3)** 805-812.

[17] Brown M K 1991 On quantization of noisy signals *IEEE Trans. Sig. Proc* 39(4) 836-841.

[18] Widrow B 1956 A study of rough amplitude quantization by means of Nyquist sampling theory *IRE Trans. Circuit Theory* **3(4)** 266-276.

[19] Widrow B Kollar I and Liu M-C 1996 Statistical theory of quantization *IEEE Trans. Instrum. and Meas* **45(2)** 353-361.

[20] Bennett W R 1948 Spectra of quantized signals *Bell Syst. Tech. J.* **27** 446-472.

Journal of Physics: Conference Series **457** (2013) 012005 doi:10.1088/1742-6596/457/1/012005

Modelling ultrasonic array signals in multilayer anisotropic materials using the angular spectrum decomposition of plane wave responses

Yousif Humeida, Valerie J Pinfield, and Richard E Challis

Electrical Systems and Applied Optics Division, Faculty of Engineering, University of Nottingham, University Park, Nottingham NG7 2RD

Yousif.humeida@nottingham.ac.uk

Abstract. Ultrasonic arrays have seen increasing use for the characterisation of composite materials. In this paper, ultrasonic wave propagation in multilayer anisotropic materials has been modelled using plane wave and angular spectrum decomposition techniques. Different matrix techniques, such as the stiffness matrix method and the transfer matrix method, are used to calculate the reflection and transmission coefficients of ultrasonic plane waves in the considered media. Then, an angular decomposition technique is used to derive the bounded beams from finite-width ultrasonic array elements from the plane wave responses calculated earlier. This model is considered to be an analytical exact solution for the problem; hence the diffraction of waves in such composite materials can be calculated for different incident angles for a very wide range of frequencies. This model is validated against experimental measurements using the Full-Matrix Capture (FMC) of array data in both a homogeneous isotropic material, i.e. aluminium, and an inhomogeneous multilayer anisotropic material, i.e. a carbon fibre reinforced composite.

1. Introduction

Fibre reinforced composites (FRC) are being used extensively for a wide range of applications; that is mainly because of their high strength and stiffness-to-density ratios and their reduced maintenance and life cycle costs. As non-destructive testing (NDT) techniques for examining such materials have been stretched to the limits, a need to develop new techniques to test and examine their integrity and reliability has emerged. Ultrasonic testing is the most commonly used and widely successful technique in the industry for testing composites for reasons of cost, safety, and ease of deployment.

Ultrasonic phased arrays have recently been replacing the use of conventional single element ultrasonic transducers in many applications. That is largely attributed to the advantages that ultrasonic arrays offer such as the potential for better sensitivity and coverage area. Holmes et al [1] and Drinkwater and Wilcox [2] have been working on the development of array techniques for NDE applications. Their philosophy was to record a full set of all time-domain data from all the different elements of the array. This set of data, which is referred to as the Full Matrix Capture (FMC), is utilized for offline post-processing. Furthermore, they demonstrated that, by testing and assessing different types of post-processing techniques, an algorithm called the Total Focusing Method (TFM) produces the highest possible imaging resolution for inclusions and scatterers in some materials, such as metals. It is expected that by using altered versions of the above imaging algorithms for testing thick-section composite materials, significant results will be achieved. However, there is a need for a

Content from this work may be used under the terms of the Creative Commons Attribution 3.0 licence. Any further distribution of this work must maintain attribution to the author(s) and the title of the work, journal citation and DOI.
Published under licence by IOP Publishing Ltd

forward model to verify the developed algorithms. Modelling ultrasonic signals in composite materials, compared to that in metals, is a very difficult and complicated task. Composite materials, such as FRP, usually have very complex structure and they are generally modelled as inhomogeneous and anisotropic multilayer assemblies.

Plane waves in multilayer media can be simulated using techniques such as the transfer matrix method [3, 4], the global matrix method [5], and the stiffness matrix method [6]. These theories are used to study the propagation of plane ultrasound waves of infinite lateral extent in multilayer materials. They do not accurately reflect real scenarios where the transducers used for transmitting/receiving the ultrasonic waves are of finite dimensions and hence generate bounded beam acoustic signals. Therefore, it is necessary to develop techniques for modelling the response of transducers of finite dimensions. Different methods can be used to model diffraction of bounded beams; the Angular Spectrum method is one of them. It states that the beams at any finite aperture can be described as an infinite set of plane waves propagating in different directions in the spatial frequency domain. These plane waves can be used to evaluate the bounded beams from the finite aperture or transducer at any point [7, 8]. As an example, Rehman et al [9] used the angular decomposition of plane waves to calculate the reflection profile of bounded beams into anisotropic multilayer media.

In this paper, ultrasonic wave propagation in multilayer anisotropic materials has been modelled using plane wave and angular spectrum decomposition techniques. Different matrix techniques, such as the stiffness matrix method and the transfer matrix method, are used to calculate the reflection and transmission coefficients of ultrasonic plane waves in the considered media. Then, the angular decomposition technique is used to derive the diffraction of bounded waves. The current model builds on earlier work by introducing multiple transducer transmitter/receiver elements to simulate a (1D) transducer array, producing predictions of the time-domain signals at receiver elements in various transmit modes. This model is validated against experimental measurements using the Full-Matrix Capture (FMC) of array data in both a homogeneous isotropic material, i.e. aluminium, and an inhomogeneous multilayer anisotropic material, i.e. a carbon fibre reinforced composite.

2. Modelling plane waves in multilayer media

Let us assume that a bulk wave is already excited and travelling indefinitely in an infinite anisotropic material in an arbitrary direction $\vec{n}$. In this case, the equation for elastic motion will be [10]:

$$\rho \frac{\delta^2 u_i}{\delta t^2} = c_{ijkl} \frac{\delta^2 u_k}{\delta x_j \delta x_l} \tag{1}$$

Where u_i is the displacement vector, ρ is the density, and c_{ijkl} are the components of the stiffness tensor of the anisotropic material. A general solution for this wave equation can be given by the following monochromic plane wave displacements [10]:

$$u_i = P_i e^{\sqrt{-1}k(n_j x_j - vt)} = P_i e^{\sqrt{-1}(k_j x_j - \omega t)} \tag{2}$$

where P_i is the component of the displacement amplitude vector (the polarization vector), n_j is the component of the propagation direction (unit vector), and k and v are the scalar wavenumber and phase velocity, respectively. $v = \omega/k$, where ω is the circular frequency. $k_j = kn_j$ are the components of the vector wavenumber.The Christoffel equation can be used to calculate the different values of wavenumbers and phase velocities in the above equation and hence the different values of the displacements and stresses can be obtained.

If a bulk wave is travelling in an anisotropic medium (medium 1), interrupted by an interface, and then continues to travel in a different anisotropic medium (medium 2), as can be seen in Figure 1, *Snell's law* and the *continuity conditions* will govern the behaviour of the wave. Three different waves, one quasi-longitudinal, and two quasi-shear, will refract and start travelling in medium 2 and three similar waves will reflect back and travel in medium 1. The *Snell's law* states that: if a plane wave is travelling from medium 1 to medium 2 through a perfect interface, firstly, the entire incident, refracted, and reflected waves will propagate in the same plane, parallel to the x_1 axis in this case. Secondly, all the projections of the wavenumbers of all these waves on the interface will be identical [10], i.e.

$$\left[k^{(I)} \sin\theta^{(I)}\right]^1 = \left[k^{(R)} \sin\theta^{(R)}\right]^1 = \left[k^{(T)} \sin\theta^{(T)}\right]^2 \qquad 3$$

Generally if the interface or the boundary is perfect and is bounding two solid media 1 and 2, the *continuity condition* states that the displacement vector u_i and the stress vector normal to the interface σ_{i1} have to be equal in the two sides of the interface. Yet for some special cases, some of the continuity conditions might not be applicable anymore. For example if one of the media is a nonviscous fluid, like water, it is found that the shear stresses vanish and the tangential displacements are not necessarily equal anymore. More information about bulk waves in anisotropic materials can be found in [11].

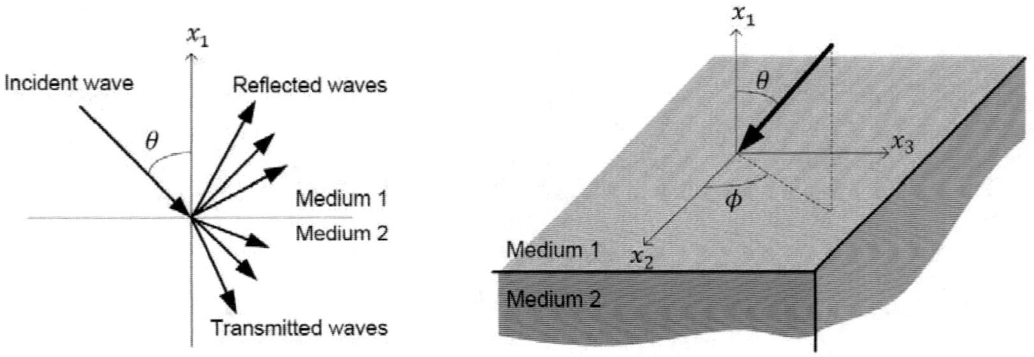

Figure 1. A plane wave in a finite anisotropic material

To study plane waves in multilayer materials, a plate consisting of an arbitrary number $n+1$ of different layers which are perfectly bonded and stacked normal to the x_1 axis (parallel to the $x_2 - x_3$ plane), as shown in Figure 2, is assumed. This system consists of $(n-1)$ different layers which are bounded with two semi-infinite layers (layer 0) at the top and (layer n) at the bottom. Each layer is assumed to have infinite length in both x_2 and x_3 directions and a finite thickness of d_k in the x_1 direction. A plane wave is assumed to be travelling in the upper half space towards the multilayer system in a propagation direction $\vec{n}$. This wave is assumed to enter the system at the origin of the global coordinate axes having an incident angle of θ with the x_1 axis and its wave vector projection on the $x_2 - x_3$ plane is assumed to have an arbitrary angle of ϕ with the x_2 axis, all illustrated in Figure 2. In this case, two of the three different k_2 and k_3 wavenumber components will already be known from the semi-infinite material by applying Snell's law and this equation will only be used to get the remaining wavenumber k_1 [12]. The problem of the wave propagation in anisotropic multilayer systems, such as the one shown in Figure 2, will be tackled at this point using two techniques: the transfer matrix method and the stiffness matrix method [6].

In the transfer matrix method, the stress and displacement at the bottom of any arbitrary layer m can be related to the stress and displacements at the top of that layer as follows:

$$\begin{bmatrix} u_i^m(d_m) \\ \sigma_{1i}^m(d_m) \end{bmatrix} = T_m \cdot \begin{bmatrix} u_i^m(0) \\ \sigma_{1i}^m(0) \end{bmatrix} \quad 4$$

where T_m is the Transfer matrix for the layer m. Iterative relations can be used to relate the stresses and displacements at the top of the whole plate to those at its bottom.

$$\begin{bmatrix} u_i^{n-1}(d_{n-1}) \\ \sigma_{1i}^{n-1}(d_{n-1}) \end{bmatrix} = T_{n-1} \cdot T_{n-2} \cdot \ldots \cdot T_1 \begin{bmatrix} u_i^1(0) \\ \sigma_{1i}^1(0) \end{bmatrix} \quad 5$$

On the other hand, if the stresses at the top and bottom of any specific layer have been related to the displacements at the top and bottom of that layer, a stiffness matrix is derived as follows:

$$\begin{bmatrix} \sigma_{1i}^m(0) \\ \sigma_{1i}^m(d_m) \end{bmatrix} = S_m \cdot \begin{bmatrix} u_i^m(0) \\ u_i^m(d_m) \end{bmatrix} \quad 6$$

Where S^m is the Stiffness matrix in layer m.

Now a recursive technique can be used in order to calculate the total stiffness matrix that relates stresses and displacements at the top and bottom of a whole system of multilayer media.

Either of the transfer matrix or the stiffness matrix can be used to calculate the transmission and reflection coefficients for the whole multilayer medium.

The plane wave model has been implemented and provided to us by Prof Michel Castaings (I2M, Universite de Bordeaux I, France).

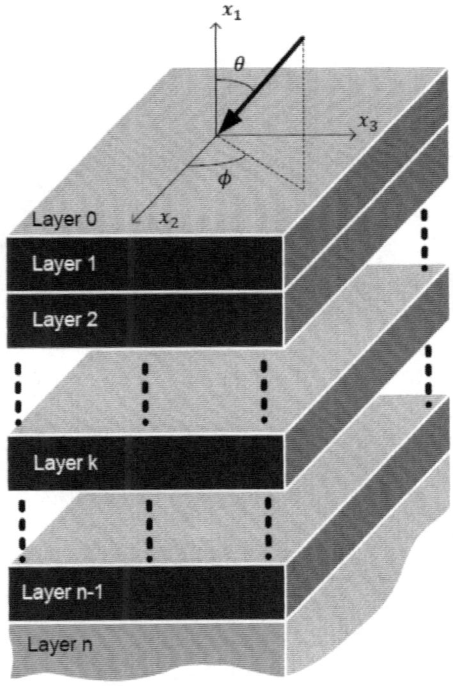

Figure 2. Plane wave with an arbitrary direction incident into the system of $n + 1$ different layers

3. Modelling bounded acoustic beams

A finite dimension transducer is used to transmit monochromatic acoustic waves in an infinite inviscid liquid where the face of the transducer is normal to the x_1 axis as shown in Figure 3. The transducer is assumed to be infinitely long in the x_2 direction while having a finite width of W in the x_3 axis direction. The finite transducer generates a longitudinal acoustic wave with a field $f(x_1 = 0, x_3, \omega)$ at its face that has a propagation vector $\vec{n}$ normal to the face of the transducer.

The angular spectrum method proposes that the Fourier transform can be used to generate the function $F(k_3, \omega)$ as follows [7]:

$$F(k_3, \omega) = \int_{-\infty}^{\infty} f(0, x_3, \omega)\ e^{-jk_3x_3}.dx_3 \qquad 7$$

$F(k_3, \omega)$ can be interpreted as an infinite number of plane waves with different values of axial wavenumbers k_3 and describes the acoustic signal from the finite transducer in the spatial frequency domain.

Now the inverse Fourier transform can be used to obtain the space varying field at any point as shown by the following equation:

$$f(x_1 = -L, x_3, \omega) = \int_{-\infty}^{\infty} F(k_3, \omega)\ e^{j(-k_1L+k_3x_3)}.dk_3 \qquad 8$$

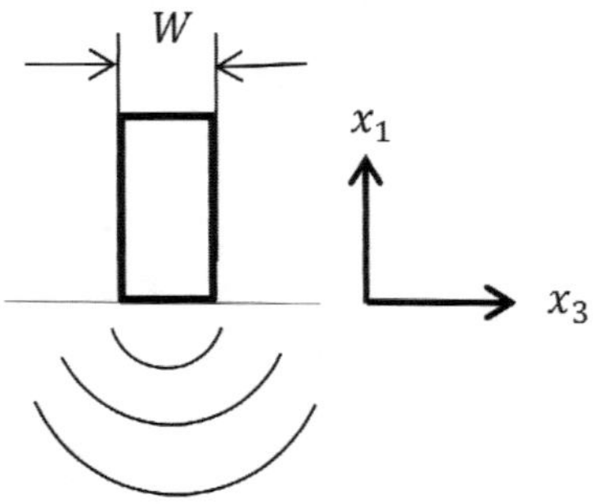

Figure 3. Finite dimension transducer transmitting bounded acoustic waves in an infinite inviscid liquid, the long axis of the transducer element is parallel to the $\boldsymbol{x_2}$ axis

For the purposes of validation, the results of the Angular Spectrum model were compared to the analytical solution of the Rayleigh-Sommerfeld integral which describes the problem of wave propagation of bounded beams in 2-D. More details for the derivation of this integral can be found in [7, 8]. The integral is given as follows:

$$f(x_1, x_3) = \frac{1}{j\lambda}\int_W f(x_1 = 0, x_3)\frac{e^{j\vec{k}.\vec{r}}}{\vec{r}} \cos\theta\,.dx_3 \qquad 9$$

4. Modelling ultrasonic array signals in different materials

The plane wave and the angular spectrum models described in previous sections can be used to model and simulate the time domain pulse echo signal transmitted by transducer 1 into a multilayer material and received by transducer 2 as can be seen in Figure 4

The plane wave model is used to calculate the fields reflected back from the material for a set of different frequencies and wavenumbers. This spectrum is then multiplied by directivity functions and

transducer transfer function in order to account for the beam spread and time domain impulse response for the two transducers, respectively. Then, the 2-D Fourier transform is used to convert these signals from the frequency and wavenumber domains to the time and space domains.
The directivity function D of a transducer of width W can be calculated analytically using equation 10 [1] whereas and for the purposes of this study transducer impulse responses has been determined experimentally.

$$D = sinc\left(\frac{\pi W \sin\theta}{\lambda}\right) \quad 10$$

The same method can be used to simulate the ultrasonic response from the different combinations of transmit and receive elements of a 1-dimensional linear array coupled to a multilayer medium. Since most linear arrays consist of identical elements, a minimal number of calculations are needed to model a whole set of Full Matrix Capture data for these arrays. More details about this model can be found in [13].

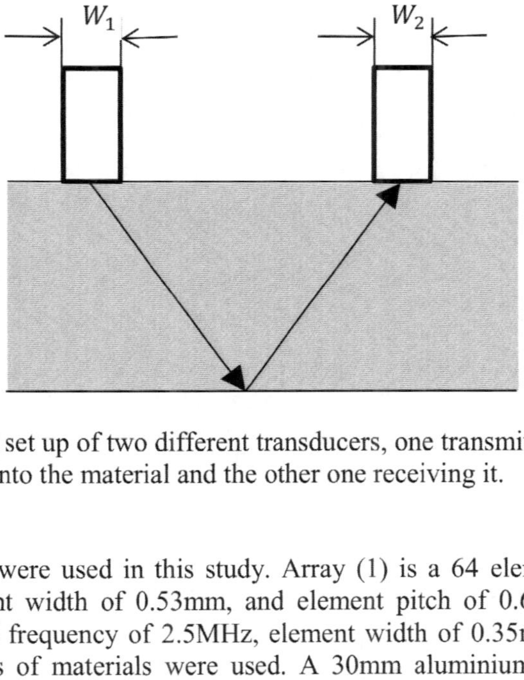

Figure 4. The experimental set up of two different transducers, one transmitting the ultrasonic signal into the material and the other one receiving it.

5. Materials and methods

Two different linear arrays were used in this study. Array (1) is a 64 element array with a central frequency of 5MHz, element width of 0.53mm, and element pitch of 0.63mm. Array (2) is a 64 element array with a central frequency of 2.5MHz, element width of 0.35mm, and element pitch of 0.5mm. Two different types of materials were used. A 30mm aluminium plate with a density of 2780kg/m^3, compressional wave velocity of 6290m/s, and a shear wave velocity of 3000m/s and a 16mm 128 layer carbon fibre reinforced composite with fibre orientations of (0°,45°,90°,-45°), repeated. Properties of individual layers in the composite can be found in [12]. Water has been used throughout all the simulation calculations as a semi-infinite bounding layer with a density of 1000kg/m^3 and compressional wave velocity of 1500m/s.
The model has been programmed using MATLAB. For all simulations presented in this paper sampling rates of 10MHz and 1mm^{-1} in the frequency and wavenumber domains, respectively, have been used.

6. Results

Figure 5 shows results for the calculation of reflection coefficients of a 6 MHz ultrasonic signal on an aluminium plate immersed in water. These results show instabilities in the transfer matrix method for the angles 23 up to 33 degrees whereas stable results are obtained for the stiffness matrix method and confirming earlier findings, by other authors such as [14] and [15], that the stiffness matrix method is more efficient and stable than the transfer matrix method.

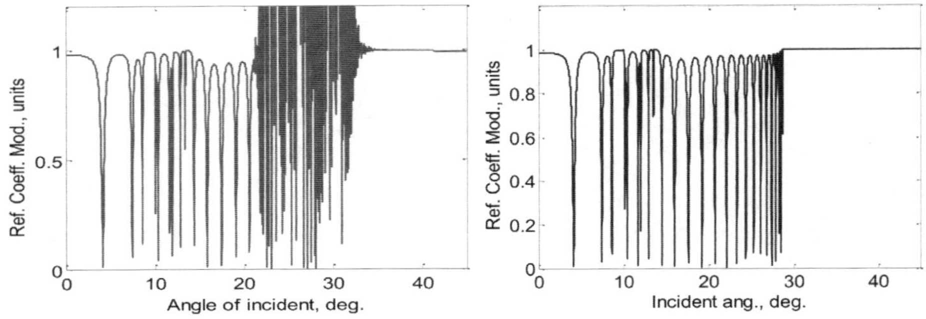

Figure 5. Reflection coefficient of a 6 MHz ultrasonic signal on an aluminium plate immersed in water using the transfer matrix method (top left) and the stiffness matrix method (top right).

Figure 6 compares the pressure profile in water at the x_1 axis (i.e. $x_2 = 0$) calculated by the Angular Spectrum method and the analytical model. The results in Figure 6 suggest that there is excellent agreement between the Angular Spectrum model and the analytical solution and hence the Angular Spectrum method has been chosen as an adequate method to model diffraction of bounded beams in multilayer materials.

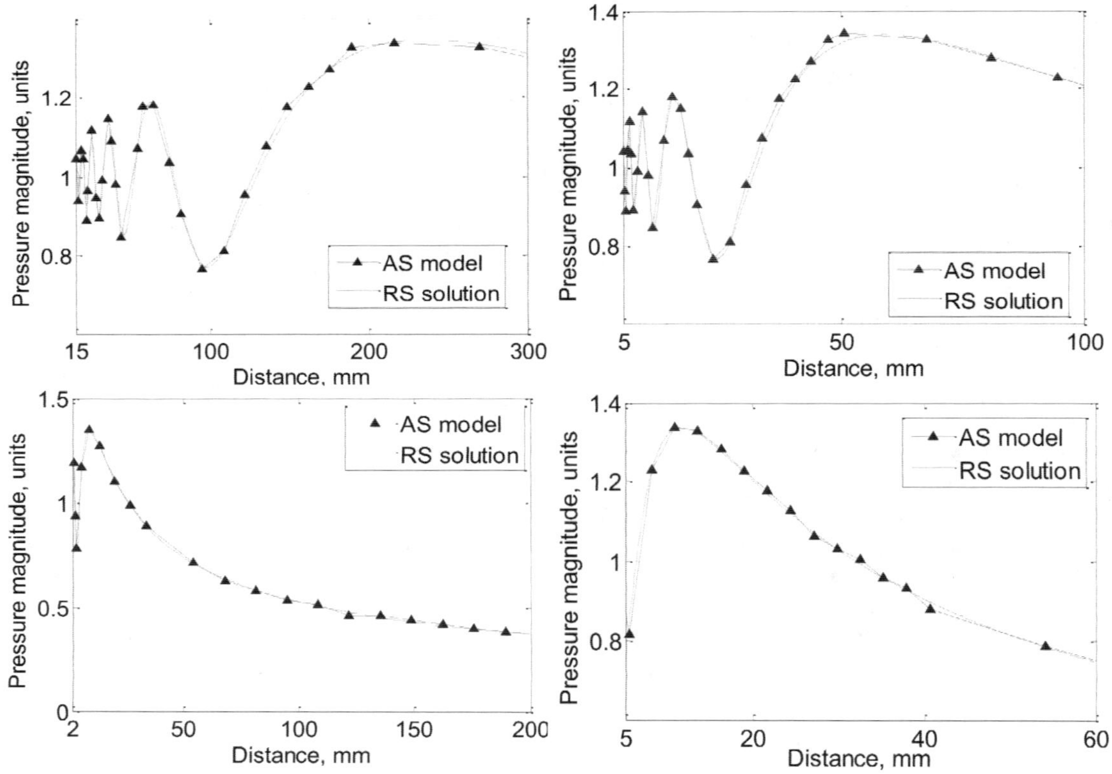

Figure 6. Comparison between the Angular Spectrum (AS) model and the Rayleigh-Sommerfeld (RS) solution for the pressure magnitude in water at the x_1 axis. The pressure was calculated for the following transducer width and frequency values: 10mm, 10MHz (top left), 10mm, 2MHz (top right), 5mm, 10MHz (bottom left), 5mm, 2MHz (bottom right).

The FMC data of array (1) has been calculated for the aluminium plate. Simulations have been compared with experimental measurements as can be seen in Figure 7, which shows signals transmitted by element 1 in the array and received by element 1, 9, 17, 25, 33, 41, 49, and 57. The first pulse that arrives back at each element is the longitudinal wave reflection from the back wall of the aluminium and because the travelled distances increase as we start receiving from the elements further from the transmitting element (element 1), the times of arrival for these pulses increase as well. It is also clear that mode converted signals arrive after the initial back wall echoes, especially for signals received by elements far away from the transmitting transducer. These can be seen from the second set of arrival pulses which are either sent as compressional waves and reflected as shear waves or sent as shear waves and reflected back as compressional waves. Such waves are not present in normal incident signals where mode conversions do not occur.

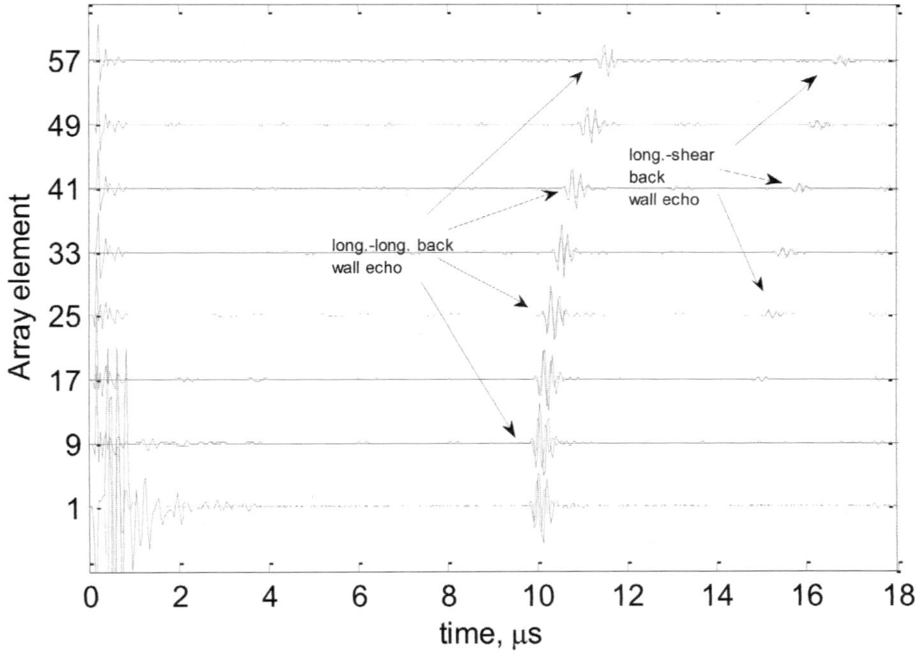

Figure 7. FMC in aluminium calculated using model (red) and experiment (blue)

The FMC data of array (2) has been calculated for the composite plate. This time, signals have been sent from element 1 and received by elements 2, 4, 6, 8, 10, 12, 14, 16, 18, and 20. Simulations have been compared with experimental measurements as can be seen in Figure 8. Two main characteristics in these signals are clear from the figure, namely the back wall echo signals and the back scattered signals from inner interfaces between the top and bottom surfaces. Attenuation in the composite sample is much higher than the aluminium sample; hence multiple reverberations of the back wall echo cannot be detected using these array signals. Both Figure 7 and Figure 8 show good agreement between the simulations and experiments when both pulse amplitudes and times of arrivals are compared. The small discrepancies in both can be attributed to the use of potentially inaccurate material properties for both samples (aluminium and composite) and possibly for bounding materials where gel has been used for the experiments yet water was considered in the simulations. Another source of discrepancies and inconsistency between the experiments and simulations might be the experimental set up itself where very thin gel layers are used to couple the ultrasonic signals into the materials, causing transducer beam deformation and the introduction of some shear waves.

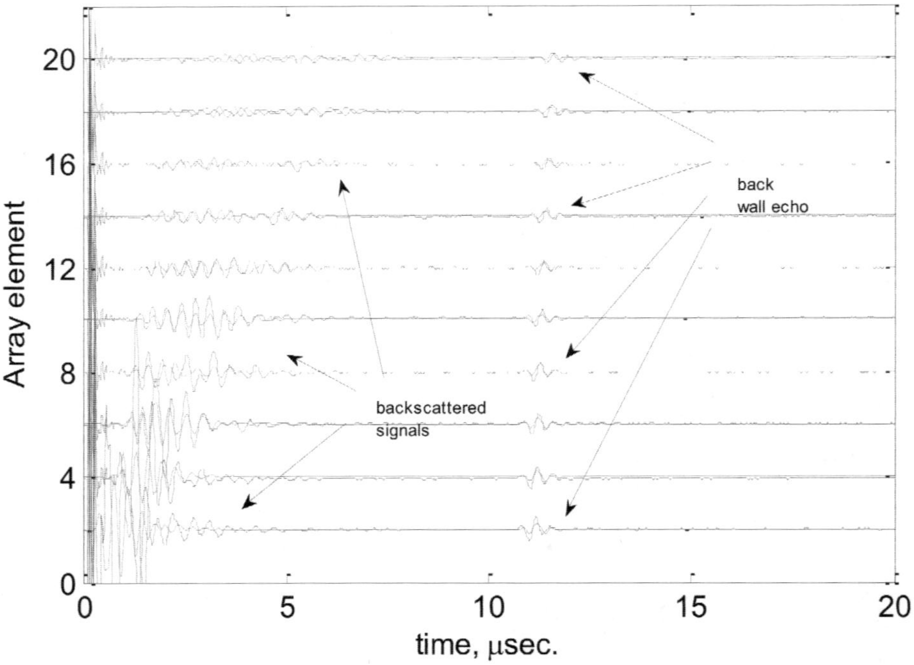

Figure 8. FMC in composite calculated using model (red) and experiment (blue)

The group velocity of the longitudinal wave in the aluminium and composite samples can be calculated, using simulations and/or experiments, from the time measurements of the back wall echo of the FMC signals from Figure 7 and Figure 8 as can be seen in Figure 9.

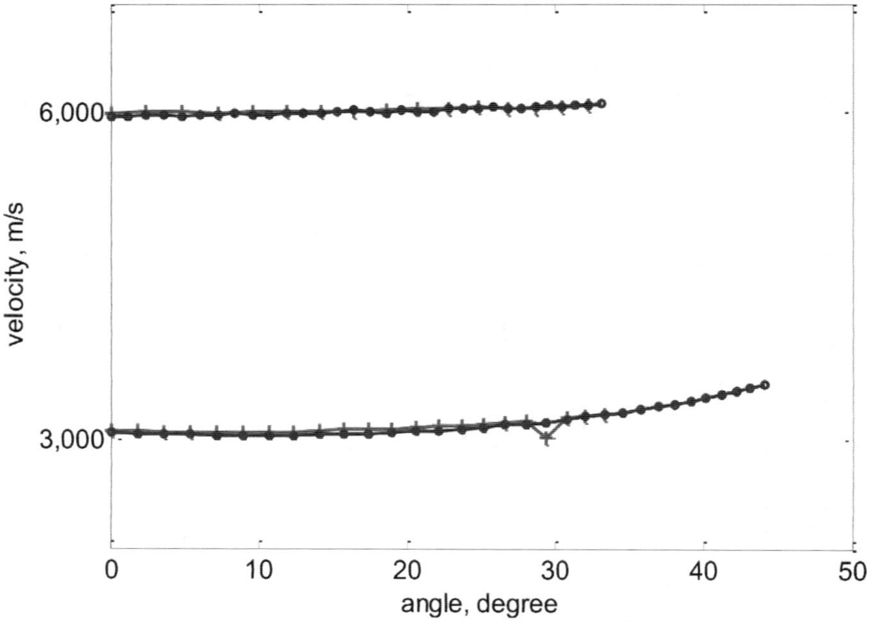

Figure 9. Longitudinal group velocity in aluminium (top curves) and composite (bottom curves) calculated from simulated (red) and experimental (blue) FMC data

7. Conclusion

The motivation behind this work was to develop a model for the propagation of ultrasonic array signals in multilayer anisotropic media. It has been demonstrated in this paper that the plane wave and angular spectrum modelling techniques not only provide a simple way of modelling the problem but also give very accurate results as shown in section 4. When multilayer systems such as FRCs are modelled using these techniques not only back wall echoes are simulated accurately but also backscattered signals from the layers in-between the front and the back wall. Such structural noise has to be considered especially when quantifying and analysing the quality of imaging techniques such as the TFM when they are used for such multilayer materials.

Only simulations in materials with planar surfaces and interfaces have been covered in this work. However, because of the accurate results that have been achieved, it is expected that this model will have significant potential in modelling and quantifying the effect of localised defects embedded within the body of a multilayer material.

8. Acknowledgement

Thanks are due to Prof Michel Castaings (I2M, Universite de Bordeaux I, France) for providing the plane wave model and to Dr Chuan Li and Prof Paul Wilcox (Ultrasonics and NDT Group, Mechanical Engineering Department, University of Bristol, UK) for providing ultrasonic array experimental data.

References

[1] Holmes C, Drinkwater B W and Wilcox P D 2005 Post-processing of the full matrix of ultrasonic transmit-receive array data for non-destructive evaluation *Ndt&E Int* **38** 701-11

[2] Drinkwater B W and Wilcox P D 2006 Ultrasonic arrays for non-destructive evaluation: A review *Ndt&E Int* **39** 525-41

[3] Thompson W T 1950 Transmission of elastic waves through a stratified solid medium *J Appl Phys* **21** 89-93

[4] Haskell N A 1953 The dispersion of surface waves on multilayered media *Bull. Seism. Soc. Am.* **43** 17-34

[5] Knopoff L 1964 A matrix method for elastic wave problems *Bull. Seism. Soc. Am.* **54** 431-8

[6] Rokhlin S I and Wang L 2002 Ultrasonic waves in layered anisotropic media: characterization of multidirectional composites *Int J Solids Struct* **39** 4133–49

[7] Goodman J W 1968 *Introduction to Fourier Optics* (Singapore: McGraw-Hill)

[8] Williams E G 1999 *Fourier Acoustics: Sound Radiation and Nearfield Acoustical Holography* (London: Academic Press)

[9] Rehman A U, Potel C and de Belleval J F 1998 Numerical modeling of the effects on reflected acoustic field for the changes in internal layer orientation of a composite *Ultrasonics* **36**

[10] Auld B A 1990 *Acoustic fields and waves in solids* (Malabar, Florida: Robert E Krierger)

[11] Hosten B 1991 Reflection and Transmission of Acoustic Plane-Waves on an Immersed Orthotropic and Viscoelastic Solid Layer *J. Acoust. Soc. Am.* **89** 2745-52

[12] Hosten B and Castaings M 1993 Transfer-Matrix of Multilayered Absorbing and Anisotropic Media - Measurements and Simulations of Ultrasonic Wave-Propagation through Composite-Materials *J. Acoust. Soc. Am.* **94** 1488-95

[13] Humeida Y, Pinfield V J and Challis R E 2012
Simulation of transducer array measurements on composite materials with defects *in preparation*

[14] Pialucha 1992 The Reflection Coefficient From Interface Layers in NDT of Adhesive Joints. In: *Mechanical Engineering*: Imperial College London)

[15] Wang L and Rokhlin S I 2001 Stable reformulation of transfer matrix method for wave propagation in layered anisotropic media *Ultrasonics* **39** 413-24

Resonant ultrasonic attenuation in emulsions

Thomas Brunet, Benoit Mascaro, Olivier Poncelet, Christophe Aristégui
Univ. Bordeaux, CNRS, I2M UMR 5295, F-33405 Talence, FRANCE

Simon Raffy, Olivier Mondain-Monval
Univ. Bordeaux, CNRS, CRPP UMR 5295, F-33600 Pessac, FRANCE

Jacques Leng
Univ. Bordeaux, CNRS, LOF UMR 5258, F-33608 Pessac, FRANCE

E-mail: thomas.brunet@u-bordeaux1.fr

Abstract. We report the achievement of scattering resonant emulsions devoted to the frequency-control of acoustic attenuation in the megahertz range. By means of robotics, we produced highly monodisperse, in both size and shape, fluorinated-oil droplet suspensions, providing experimental evidence of several Mie scattering resonances. Ultrasonic experiments performed in such complex media are compared, with an excellent quantitative agreement, to theoretical predictions derived within the framework of the independent scattering approximation.

1. Introduction

In disordered scattering media, the wave attenuation is basically due to i) the absorption within the surrounding media and ii) the wave scattering by the inclusions. To enhance the attenuation, an alternative is to exploit the scattering resonances of the inclusions, which depend on both their shape and their physical properties. In the audible domain ($f \leq 20$ kHz), many efforts have been devoted to the achievement of low-frequency resonators such as core-shell particles [1], or slip hollow spheres [2]. For higher frequencies ($f \geq 50$ kHz), the manufacture of such sophisticated objects is a difficult task due to the sub-millimetric scale required in the ultrasound domain. Despite their short lifetimes, the air-bubbles turned out to be an efficient way for ultrasonic insulation [3, 4]. The low-frequency resonators are not the only promising way to enhance acoustic attenuation: Mie scattering resonances could be a feasible alternative, provided a high sound speed contrast between the scatterers and the surrounding medium [5]. Although resonant scattering from spherical cavities in viscoelastic media has theoretically been considered for many decades [6], no experimental results have been reported about *resonant emulsions*, due to the coexistence of multiresonances and the difficulties associated with observing them at the high frequencies required [7]. Up to now, such random mixtures have only been considered in the long wavelength limit [8]. In this paper, we focus our attention on the resonant scattering regime occurring in emulsions.

Content from this work may be used under the terms of the Creative Commons Attribution 3.0 licence. Any further distribution of this work must maintain attribution to the author(s) and the title of the work, journal citation and DOI.
Published under licence by IOP Publishing Ltd

2. Materials and methods

2.1. Sample fabrication

To observe Mie scattering resonances, we used a low absorption suspending gel that behaves almost like water in the megahertz range we investigated (v_0 = 1490 m.s^{-1}, ρ_0 = 1 g.mm^{-3} and $\alpha_0 = 10^{-4}$ MHz^{-2}.mm^{-1}). The yield stress of this water-based gel ($\sim$ 10 Pa) circumvents possible particle sedimentation but does not affect the wave propagation, *i.e.*, no shear wave can propagate into such fluid-like matrix. To enhance Mie scattering resonances, high compressibility and mass density contrasts between the two phases are required. As suggested in a recent numerical study [9], we used the fluorinated oil FC 40 (Fluorinert ®) possessing a rather low sound speed for a pure liquid ($v_1 \simeq$ 640 m.s^{-1}) and a high mass density (ρ_1 = 1.85 g.mm^{-3}). As previously done for bubbly media [10], the emulsion was achieved thanks to robotics [11]. The fluorinated oil was continuously injected within the gel matrix by means of a moving syringe. Typically, 100 μm-radius droplets were obtained with large displacement velocities ($\sim$ 100 mm.s^{-1}) and low rates of oil-flow ($\sim$ 10 μL.min^{-1}). On one hand, this immiscible-oil-in-water emulsion is much more stable than a bubbly medium that makes its conservation longer. In addition, we obtained an excellent monodispersity (1%-polydispersity) as shown in Fig. 1. The volume fraction Φ_v of the fluorinated oil was estimated from the liquid quantity injected in the gel matrix. Then, the emulsion was optically characterized by the image analysis of about a hundred of droplets, of which radius measurements were made with a 1 μm-accuracy (see the particle size distribution in Fig.1). In this work, the mean radius of such emulsion made of fluorinated-oil droplets is $<a>$ = 104.5 μm and the volume fraction is Φ_v = 0.5%. Note that the disorder arises from the emulsion transfert into a cell devoted to acoustic measurements

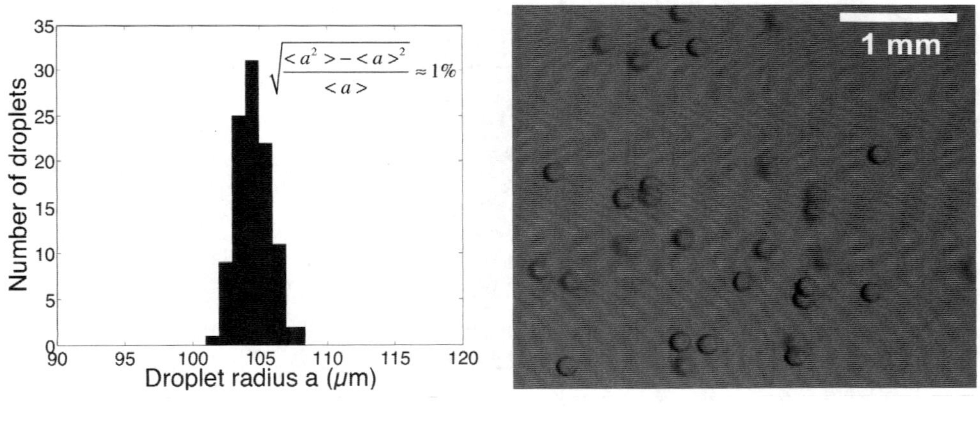

Figure 1. Discrete droplet-radius distribution measured by optical microscopy of an emulsion of fluorinated-oil droplets with $<a>$ = 104.5 μm and Φ_v = 0.5%.

2.2. Acoustic measurements

Ultrasonic measurements were carried out in a large water tank (80x40x30 cm^3) by multi-echo spectroscopy [12]. An ultrasound pulse was generated by a broadband pulser/receiver connected to a broadband longitudinal wave transducer, having a central frequency of 10 MHz. The samples were poured into a polystyrene rectangular cell of which the width d = 15 mm was large enough to separate the multiple echoes on the walls. By means of this reflexion technique, we first measured the complex-valued wavenumber $k_0 = k_0' + ik_0''$ of the pure gel-matrix ($k_0' = \frac{\omega}{v_0}$ and $k_0'' = \alpha_0$ with $\omega = 2\pi f$ standing for the angular frequency). Then, we measured the effective attenuation coefficient of the emulsion that exhibits several sharp peaks (see Fig. 2).

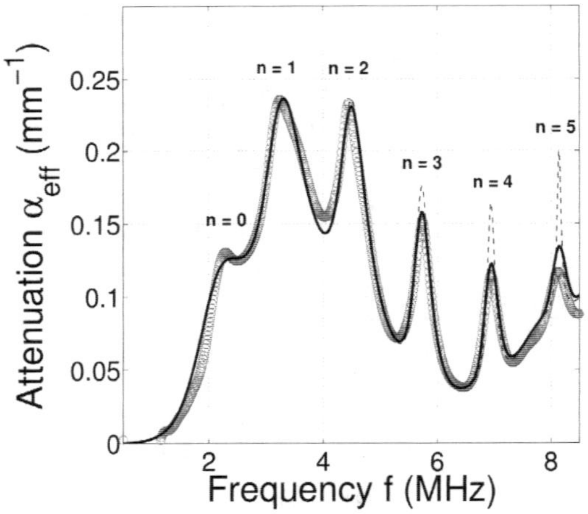

Figure 2. Acoustic attenuation measured in an emulsion made of fluorinated-oil droplets with $< a > = 104.5$ μm and $\Phi_v = 0.5\%$ (red open circles). Theoretical predictions calculated from the ISA model for a monodisperse emulsion (dashed line) and for a polydisperse emulsion (solid line) taking into account the particle size distribution measured by optical microscopy (Fig. 1).

3. Results and discussions

Since the volume fraction of oil droplets is quite low ($\Phi_v \leq 1\%$), we analyzed the experimental results within the framework of the independent scattering approximation (ISA), which is equivalent to the Foldy's model [13]. For polydisperse media [14], the effective wavenumber k_{eff} of the coherent pressure-wave propagating is given by:

$$k_{eff}^2 = \left(\frac{\omega}{v_{eff}} + \mathrm{i}\alpha_{eff}\right)^2 = k_0^2 + \int_a 4\pi\eta(a)\mathrm{d}a f_s(a) \qquad (1)$$

where $\eta(a)$ is the number of droplets per unit volume whose radius is between a and $a + \mathrm{d}a$ ($\Phi_v = \int_a \frac{4}{3}\pi a^3 \eta(a)\mathrm{d}a$) and $f_s = \frac{1}{\mathrm{i}k_0}\sum_{n=0}^{\infty}\{2n+1\}S_n(k_0 a)$ is the forward scattering function [15]. The complex-valued coefficients S_n, which depend on $k_0 a$ and also on the ratios $\frac{\rho_1}{\rho_0}$ and $\frac{v_1}{v_0}$, describe the scattering properties of a single droplet. The index n is associated with the nth mode, *e.g.*, n=0, n=1, n=2, n=3, n=4 and n=5 refer to the monopolar, dipolar, quadrupolar, hexapolar, octupolar and decapolar modes, respectively. The resonance frequencies of each mode fix the frequency-locations of the attenuation peaks[1]. As the absorption within the pure gel-matrix is very low ($\alpha_0 \simeq 10^{-2}$ mm^{-1} at 10 MHz), it could reasonably be neglected. In addition, Eq. 1 can be linearized since the volume fraction of droplets is quite low ($\Phi_v \leq 1\%$). Thus, the effective attenuation coefficient α_{eff} can be derived from the imaginary part of the effective wavenumber k_{eff} given in Eq. 1 as following:

$$\alpha_{eff} = \int_a \frac{-2\pi v_0^2 \eta(a)\mathrm{d}a}{\omega^2} \mathrm{Re}\left[\sum_{n=0}^{\infty}\{2n+1\}S_n(k_0 a)\right] \qquad (2)$$

From the droplet-radius distribution displayed in Fig. 1, we calculated the effective attenuation for the emulsion, revealing an good quantitative agreement with experiments, as

[1] The first-order resonances of the monopolar ($n = 0$), dipolar ($n = 1$), quadrupolar ($n = 2$), hexapolar ($n = 3$), octupolar ($n = 4$) and decapolar ($n = 5$) modes occur at 2.4, 3.3, 4.5, 5.7, 6.9 and 8.1 MHz, respectively, as shown in Fig. 2.

shown in Fig. 2. Moreover, a much better agreement is found by taking into account the particle size distribution measured by optical microscopy, rather than considering a monodisperse emulsion. The impact of polydispersity is particularly observable for high-order resonances as recently reported [14]. Thus, the ISA well-describes Mie scattering resonances occurring in dilute emulsions ($\Phi_v \leq 1\%$). As suggested by Richter *et al* [17], the high sensitivity to resonance peak frequency-positioning could make the *resonant regime* a useful tool for *particle sizing*, for which the long-wavelength regime is usually used [7].

4. Conclusion

In summary, we have presented clear experimental evidence about Mie scattering resonances occurring in dilute emulsions with high compressibility and mass density contrasts. This class of strongly scattering media might be used as absorbing materials provided the volume fraction of oil-droplets is appreciably increased ($\geq 10\%$). From a theoretical point of view, the interactions between oil-droplets should be taken into account *via* multiple-scattering theory.

References

[1] Liu Z, Zhang X, Mao Y, Zhu Y Y, Yang Z, Chan C T and Sheng P 2000 *Science* **289** 1734
[2] Ding C, Hao L and Zhao X 2010 *J. Appl. Phys.* **108** 074911
[3] Leroy V, Bretagne A, Fink M, Willaime H, Tabeling P and Tourin A 2009 *Appl. Phys. Lett.* **95** 171904
[4] Bretagne A, Tourin A and Leroy V 2011 *Appl. Phys. Lett.* **99** 221906
[5] Li J and and Chan C T 2004 *Phys. Rev. E* **70** 055602
[6] Gaunaurd G C and Uberall H 1978 *J. Acoust. Soc. Am.* **63** 1699
[7] Challis R E, Povey M J W, Mather M L and Holmes A K 2005 *Rep. Prog. Phys.* **68** 1541
[8] McClements J and Povey M J W 1989 *Journal of Physics D:Applied Physics* **22** 38
[9] Dubois J, Aristegui C, Poncelet O and Shuvalov A L 2011 *J. Phys.: Conf. Series* **269** 012004
[10] Leroy V, Strybulevych A, Page J H and Scanlon M G 2008 *J. Acoust. Soc. Am.* **123** 1931
[11] Brunet T, Raffy S, Mascaro B, Leng J, Wunenburger R, Mondain-Monval O, Poncelet O and Aristégui C 2012 *Appl. Phys. Lett.* **101** 011913
[12] Norisuye T, Sasa S, Takeda K, Kohyama M and Tran-Cong-Miyata Q 2011 *Ultrasonics* **51** 215
[13] Foldy L L 1945 *Phys. Rev.* **67** 107
[14] Mascaro B, Brunet T, Poncelet O, Aristégui C, Raffy S, Mondain-Monval O and Leng J 2013 *J. Acoust. Soc. Am.* **133** 1996
[15] Aristegui C and Angel Y C 2007 *Wave Motion* **44** 153
[16] `http://www.ondacorp.com/tecref_acoustictable.shtml`
[17] Richter A, Babick F and Stintz M 2006 *Ultrasonics* **44** e483

Journal of Physics: Conference Series **457** (2013) 012007 doi:10.1088/1742-6596/457/1/012007

Data set reduction for ultrasonic TFM imaging using the effective aperture approach and virtual sources

S Bannouf [1], S Robert [1], O Casula [1] and C Prada [2]

[1] CEA, LIST, DISC, F-91191 Gif-sur-Yvette cedex, France
[2] Institut Langevin, 10 rue Vauquelin, 75231 Paris Cedex 05, France

E-mail: souad.bannouf@cea.fr

ABSTRACT

The Total Focusing Method (TFM) is sometimes described in the literature as the "gold standard" compared to conventional imaging techniques. TFM is an algorithm that usually post-processes the full matrix of data, also called Full Matrix Capture (FMC). Real-time piloting of such an algorithm is heavy due to the large number of firings (N for a N-element array) and the large number of signals (N×N) to process that tend to decrease the frame rate and, consequently, the inspection speed. This problem can be overcome to some extent if only a few elements are activated which is equivalent to using a sparse array in transmit. The PSF (Point Spread Function) provides information about important images parameters: lateral resolution and contrast. An algorithm based on PSF optimization is proposed to obtain both the number of transmit pulses and the location of the active elements. However, reducing the number of emissions induces a loss in transmitted energy. To compensate it, each transmit pulse is carried out by multiple transmit elements that emulate a single "virtual" element. The method is evaluated on experimental data in a realistic NDT configuration by comparison of images obtained with FMC and SMC (Sparse Matrix Capture) acquisitions.

1. Introduction

Ultrasound methods are commonly used in non destructive testing (NDT), for example, in the field of nuclear energy and aerospace, where safety requirements are important. They may allow detection, location, identification and sizing of the flaws of mechanical parts and assemblies without altering their mechanical properties. Real-time inspection may prevent industry from failure, improve quality while limiting costs. Ultrasonic transducer arrays are more and more widespread for detection and imaging. The large number of possibilities offered by them has opened the development of various imaging methods. Among all the existing imaging techniques, the Total Focusing Method (TFM) is sometimes described as the "gold standard" compared to conventional imaging techniques [1], [2]. TFM is an algorithm that usually post-processes the full array response matrix, also called Full Matrix Capture (FMC), obtained by exciting each element of an ultrasonic phased array individually while reception is performed by all the elements. For an array of N elements, then N^2 signals are collected. Image acquisition speed for conventional system is determined by the number of lines in the Bscan. Reducing the number of lines by limiting the field of view and/or decreasing the lateral resolution

Content from this work may be used under the terms of the Creative Commons Attribution 3.0 licence. Any further distribution of this work must maintain attribution to the author(s) and the title of the work, journal citation and DOI.
Published under licence by IOP Publishing Ltd

results in an increase of the acquisition speed. It is not the case for TFM imaging whose acquisition speed can be increased by simply reducing the number of independent firings. This is equivalent to defining a sparse array in transmit while reception is performed by either all or some of the elements of the array.

The design of sparse arrays was intensely studied in the 1990s in the medical imaging field by various authors such as Lockwood, who put forward multiple algorithmic strategies for designing periodic sparse arrays using the effective aperture function [3]. Recently, the design of sparse linear array has been transposed to NDT [4] and is still an open question. Sparse arrays for synthetic aperture imaging can be defined in several ways such as optimizing the elements weights of a sparse array where the location of the active elements is already determined [4], [5], or optimizing the location of the active elements when weights are determined [6]. In [7], [8] the optimization is simultaneously applied for locations and weights of a fixed number of active elements. The advantage provided by sparse arrays is that they reduce the amount of data to be managed by reducing the number of transmissions. This leads to two drawbacks associated with the two main types of noise: electronic and material noise. The first drawback induced by the reduction of transmit firings is that less energy is put in and thus less signals contribute to the image at each point. Assuming that noise is mainly an uncorrelated electrical one, the loss of transmitted energy results in a loss of Signal to Noise Ratio (SNR). To compensate this loss, Karaman *et al.* suggests the use of "virtual" sources. Each of them consists in a group of several elements time delayed to transmit a cylindrical wave, thus emulating a single "virtual" element located below the array with higher acoustic power [3], [9]. The second drawback is that the sparse transmit aperture may result in a Point Spread Function (PSF) that is worse than the one provided by a full array. This can be problematic when imaging flaws in noisy parts since the sidelobes level of a PSF characterizes material noise [10]. Thus, the sparse transmit aperture may lead to a PSF that worsens the image quality because the signal to coherent noise will be worse.

The aim of this work is to reduce the data volume to be managed in TFM imaging while limiting loss of image quality. In other words, the goal is to determine the optimal Sparse Matrix Capture (SMC) providing a TFM image equivalent to that obtained by post-processing a FMC. This paper presents a method to design a sparse array providing a TFM image close to the one obtained by post-processing a FMC. Contrary to the different approaches encountered in the literature, the proposed designing method consists in finding both accurate number and best location of active elements according to the weights applied to the full array in reception. The optimization is based on the comparison of the PSF of the sparse array with the one generated with full matrix of data. The result is then evaluated in near field condition according to NDT restrictions such as structural noise. The loss in signal due to the reduced number of firings is compensated by the use of "virtual" sources.

2. Total focusing method and point spread function

The TFM imaging technique is a synthetic focusing method applied to elementary signals acquired with a transducer array. This algorithm is generally applied to a FMC acquisition. For an array of N active elements, this acquisition is composed of the N×N stored impulse inter-element responses $s_{nm}(t)$ defined as the output of the element number m when the input of the element number n is a delta impulse. Once all elementary signals $s_{nm}(t)$ are stored, the principle of the TFM algorithm is to coherently sum them for each point of a Region Of Interest (ROI) [11].

Noting $T_{nm}(x, z)$ the theoretical time of flight between a pair of element (n, m) through a point P(x, z) of the ROI (see Figure 1), the image magnitude I in this point may be written as:

$$I(x,z)=\sum_{nm} w_n^T(n)w_m^R(m)s_{nm}(T_{nm}) \tag{1}$$

where

$$T_{nm}(x,z)=\frac{R_{T_n}(x,z)+R_{R_m}(x,z)}{c} \tag{2}$$

R_{Tn} and R_{Rm} are respectively the distances between the transmitter and receiver to the imaging point and c is the ultrasonic velocity in the ROI. w_n^T and w_m^R are weights applied to the elements n and m respectively.

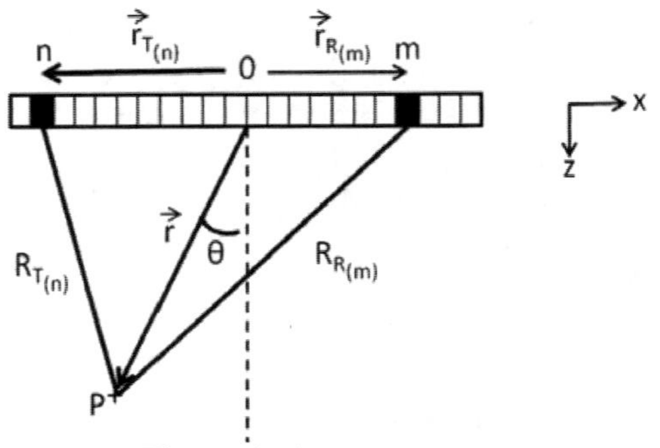

Figure 1: Array geometry

In the frequency domain equation (1) can be written as:

$$I(x,z)=\sum_{\omega}\sum_{nm} w_n^T(n)w_m^R(m)s_{nm}(\omega)d_{nm}(\omega,x,z) \quad (3)$$

where $s_{nm}(\omega)$ is the frequency spectrum of array data, and d_{nm} are focusing coefficients.

$$d_{nm}(\omega,x,z)=\exp\left[jk(R_{T_n}(x,z)+R_{R_m}(x,z))\right] \quad (4)$$

Assuming that the reflectors are located in the far-field of the array (i.e. $|\vec{r}| >> |\vec{r}_T|, |\vec{r}_R|$) and using the approximation $|\vec{r}-\vec{r}_E| \approx r - \frac{\vec{r}.\vec{r}_E}{r}$ (subscript E refers to subscripts R or T), R_T and R_R can be written as follows:

$$R_{T_n}=r-x_{T_n}.\sin\theta \quad (5)$$

and

$$R_{R_m}=r-x_{R_m}.\sin\theta \quad (6)$$

where $x_{T_n}=|\vec{r}_T|=nd$ and $x_{R_m}=|\vec{r}_R|=md$ with d the array pitch

With the previous approximation, equation (3) can be re-written as follows:

$$I(x,z)=\sum_{\omega}\sum_{nm} e^{jk2r}s_{nm}(\omega)A_{nm}(\omega,\theta) \quad (7)$$

where

$$A_{nm}\ \omega,\theta\ = w_n^T(n)w_m^R(m)e^{-jkx_{T_n}\sin\theta}e^{-jkx_{R_m}\sin\theta} \quad (8)$$

The two first terms of (7) are associated with radial resolution while A_{nm} is associated with angular resolution quality. This last one can be considered as the angular Point Spread Function (PSF) for a target in the θ direction .A Point Spread Function (PSF) corresponds to the response of an imaging system to a point-like target [4].

$$
\begin{aligned}
PSF(\omega,\theta) &= \sum_{nm} A_{nm}(\omega,\theta) \\
&= \sum_{n=0}^{N-1} w_n^T(n) e^{-jknd\sin(\theta)} \sum_{m=0}^{N-1} w_m^R(m) e^{-jkmd\sin(\theta)}
\end{aligned}
\tag{9}
$$

where λ is the wavelength.

Setting $\sin\theta = 2\pi p / Nkd$, with p denoting the frequency index, the PSF can be described as a Discrete Fourier Transform (DTF)

$$PSF(\omega,\theta) = DTF\left[w^T \otimes w^R\right] \tag{10}$$

The convolution product of the apodization functions applied in transmit (w^T) and in receive (w^R) is referred to as the effective aperture:

$$e_{TR}(n) = w^T(n) \otimes w^R(n) \tag{11}$$

Thus, the PSF can be obtained by Fourier transform of the effective aperture which represents an equivalent receive aperture that would produce an identical two-way radiation pattern if the transmit aperture was a point source [3], [7].

TFM achieves high resolution images in an extended area. It also provides finer characterization than other classical methods [12], [13]. However, the main drawback of TFM imaging is the large amount of data to acquire and to process for image reconstruction. This is time consuming and limits real-time application of the algorithm. This problem can be overcome to some extent if only a few elements are activated in transmission which is equivalent to using a sparse array in transmit. Image contrast and lateral resolution depend on the PSF characteristics. The main lobe width of the PSF determines the lateral resolution while image contrast depends on side lobe level. Thus, the relation between effective aperture and PSF can be used as a useful strategy for analysis and optimization of a sparse array. The PSF can be optimized by selecting the appropriate transmit and receive aperture functions. That is why, in the next section, the adopted method to design a sparse array is based on PSF optimization.

3. Optimization of a sparse array

3.1. Description of the optimization algorithm

The optimization of the sparse array consists in minimizing the number of transmit firings by determining both the accurate number and best locations of active elements according to the weights applied to the full array in reception.

Different algorithms have been proposed for optimization of the locations of a number of transmit elements already fixed. Most of them use one or two of the following criteria: maximum side lobe peak (SLP) below an acceptable value noted *A* and maximum main lobe width of the PSF (MLW) below a threshold *B*. In the proposed algorithm, the number of active transmit elements and their locations are determined using three criteria, those mentioned above and a third one based on the sidelobe energy (SLE) that must be less than the value denoted *C*. *A*, *B* and *C* are the different thresholds of acceptance based on characteristic of a full array radiation pattern. *A* is the maximum level value that maximum sidelobe peak is authorized to reach. *A* equals -27 dB like the higher sidelobe level measured on the PSF of a full array (reference). *B* is the threshold required for the main lobe width at -20 dB. To avoid too much loss of lateral resolution, *B* is chosen equal to 1.2 times the main lobe width of a full array radiation pattern. The maximum level of sidelobe energy (SLE), *C*, is equal to the sidelobe energy of the radiation pattern of a full array plus 3 dB.

The optimization process of an N element array can be summarized as follows:

1. Activation of the first and the last transmit elements;

Journal of Physics: Conference Series **457** (2013) 012007 doi:10.1088/1742-6596/457/1/012007

The activation of the two outer elements of the array allows maximizing the lateral resolution and speed up computation time.

2. Addition of one element to the previous sparse array;
3. Calculation of the associated radiation pattern for each possible location;
4. Determination of the best location for the added element :
   The best location is the one satisfying maximum sidelobe peak < *A* and providing the smaller main lobe width
5. If the PSF of the "new" array meets the three criteria:
   SLP < *A*,
   and MLW < *B*
   and SLE < *C*
   Then the optimized sparse array is obtained;
   Else add another element to the sparse array and repeat the operation until the criteria are satisfied.

3.2. Computation and experimental results

The previous algorithm was applied to optimize the number and locations of active transmit elements on a 2-MHz array of 58 elements of 0.8 mm pitch. Only longitudinal waves are considered and the following results are obtained by using the longitudinal velocity in ferritic steel. Depending on the aperture function applied in reception, the algorithm provides different results. Thus, when no apodization is used, the optimized sparse array is constituted of 13 active elements. When a trapezoidal windowing is applied, only 6 firings are necessary. The number of active elements is lowered when applying apodization functions in reception. Indeed, they provide the advantage of lowering side lobes (contrast improvement) but on the other side they widen the main lobe width (loss of lateral resolution) [13]. Note that the trapezoidal window was chosen because it proved to be the best compromise between loss of lateral resolution and lower sidelobes, in this configuration. Other apodization windows, such as Hamming or Blackman-Harris windows, further reduce the level of sidelobes but degrade more the lateral resolution. Figure 2 shows the comparison between the PSF of a full array with the one of a sparse array apodized in reception (Figure 2 (b)) or not (Figure 2 (a)). First, a slight increase of the main lobe width can be noted (+6 % of the FMC-PSF mainlobe width) when apodization was applied. Without apodization, the PSF of the sparse array presents sidelobes lower than the one of the full array, but locally, the sidelobe level can be higher for angles greater to 30°. When a trapezoidal apodization is used, 6 firings are sufficient to obtain a PSF which envelope fits the reference's one. Sidelobe have been attenuated by the application of the trapezoidal window to the receive elements. The use of apodization allows a good compromise between beam width, side lobe level, and number of firings.

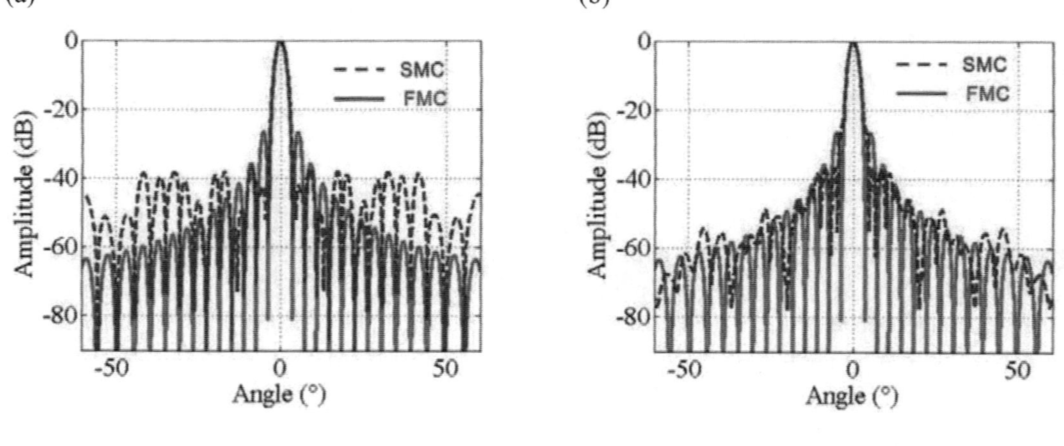

Figure 2 : Comparison of resulting PSF of a full and non apodized array and optimized sparse array. (a) : non apodized sparse array, (b) apodized sparse array (trapezoidal apodization)

The results provided by the optimization algorithm were applied to experimental data collected on a ferritic steel sample with several Side-Drilled Holes (SDH) of 2 mm diameter and located between 30 and 60 mm away from the array (Figure 3).

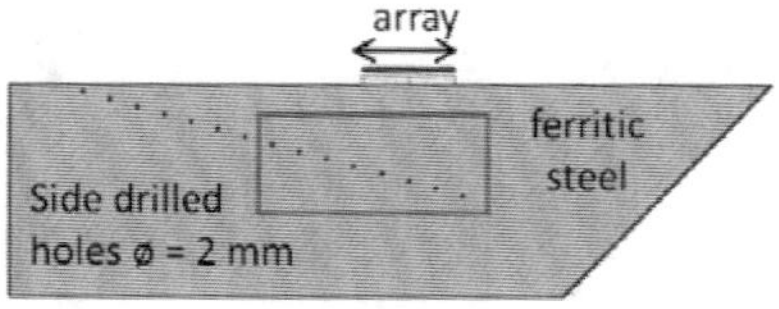

Figure 3 : Experimental configuration

The resulting images (Figure 4 and Figure 5) are obtained after TFM processing and presented on a 35 dB dynamic range. Figure 4 shows the image obtained by post-processing a FMC acquisition. It is considered as the reference image.

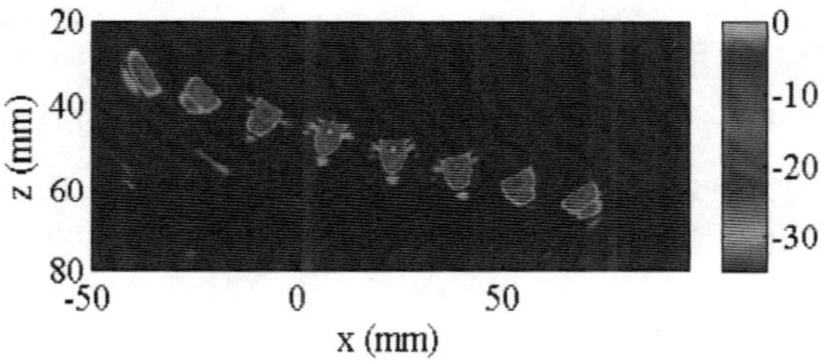

Figure 4: Image of the ferritic steel sample obtained with FMC-TFM (reference image)

Figure 5 displays four TFM images resulting of post-processing of SMC acquisitions. Figure 5(a) and Figure 5(b) were obtained without apodization and with 13 firings. The location of the active transmit elements was determined by the optimization algorithm for Figure 5(a) while random sequences were used for Figure 5(b). The sequences are reported in Table 1.

Table 1 : Numerical results provided by the optimizing algorithm and random sequences used get random SMC-TFM images

	Optimized sequences	**Random sequences**
Without apodization	1, 13, 18, 22, 24, 30, 32, 35, 38, 44, 46, 51, 58	1, 2, 5, 15, 16, 25, 37, 40, 44, 49, 49, 55, 57, 58
With apodization	1, 17, 29, 43, 44, 58	1, 16, 34, 40, 50, 58

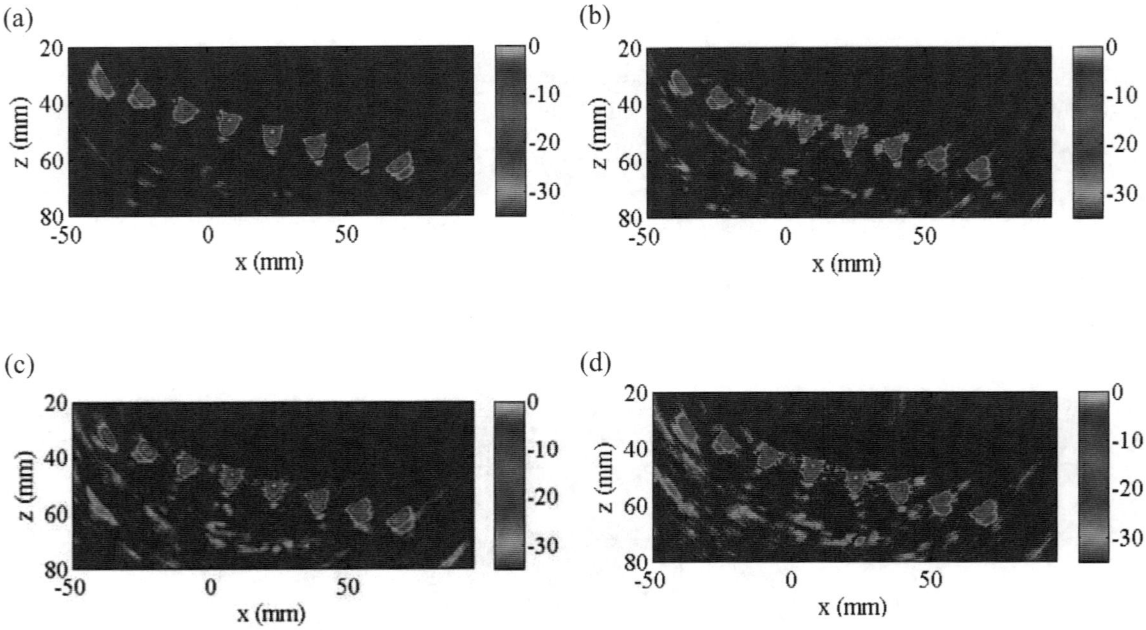

Figure 5: Images of the ferritic steel sample obtained with (a) non apodized SMC-TFM with sequences provided by the optimizing algorithm, (b) non apodized SMC-TFM with random sequences (c) trapezoidal apodized SMC-TFM with sequences provided by the optimizing algorithm, (d) trapezoidal apodized SMC-TFM with random sequences.

When comparing Figure 5(a) and Figure 5(b), it appears that image quality of the first one is better than the second one's. Indeed, it presents less artifact and sidelobe noise especially around the holes. This illustrates that the locations of the active transmit elements impacts the image quality. Transmit element location optimization is then necessary to limit image degradation due to less transmit events. This observation is also valid for Figure 5(c) and Figure 5(d) which SMC acquisitions were apodized with a trapezoidal window. It has been shown in section 3.1 that the use of a trapezoidal apodization lowers the number of necessary signals to make an image of acceptable quality. However, this results in more visible artifacts. Thus, Figure 5(c) presents more artifacts than Figure 5(a) that was obtained with a sparse matrix constituted with twice more transmit events. The extent to which artifacts appear depends on the dynamic range used for the images (Figure 6). But, in general, their amplitude is very low compared to the one of the flaw and do not perturb the detection when employing 30 dB dynamic range which is a representative of the one used in many industrial application [4].

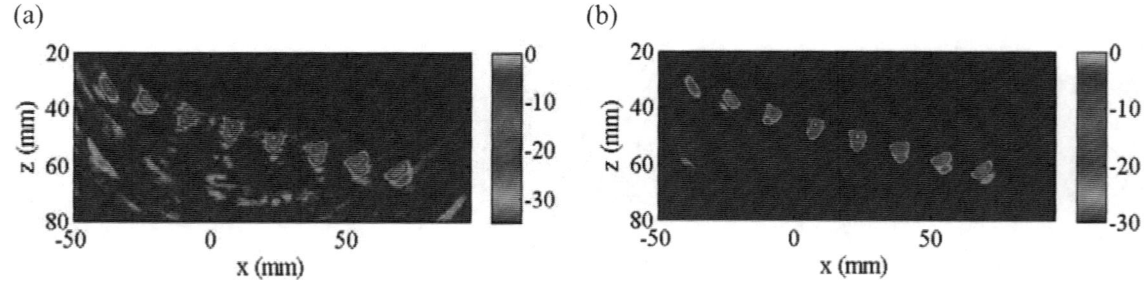

Figure 6: Sparse and apodized TFM image with (a) a 35 dB dynamic range, (b) a 30 dB dynamic range

To quantify the differences between FMC-TFM images and SMC-TFM images but also between SMC-TFM images and apodized SMC-TFM images, measurements have been performed and are reported in Table 2.

Table 2 : Quantitative characteristics of the ferritic steel sample TFM images.

	FMC-TFM image	**SMC-TFM image (no apodization)**	**SMC-TFM image (trapezoidal apodization)**
Number of firings	58	13	6
Increase of the mainlobe width (%)	-	+6%	+6%
Amplitude of hole	0 dB	-13 dB	-22 dB
SNR	45 dB	40dB	35 dB
Loss of SNR	-	-5 dB	-10 dB
Theoretical loss of SNR	-	-6 dB	-10 dB

Figure 7(a) provides the cumulative amplitude curves of TFM images of Figure 5. For a given abscissa of a TFM image, we sum the amplitudes present on all different lines of the image. In other words, each curve is obtained by summing all the lines of the corresponding TFM image. These cumulative amplitude curves and Table 2 show that as the number of firings diminishes, the measured amplitude decreases due to the reduced transmitted energy. This results in a loss of SNR which is measured on TFM images slices like the one of Figure 7(b). The SNR was calculated as the ratio of the 50mm depth SDH maximum amplitude to the average noise level taken around the SDH and at the same depth. The experimental loss of SNR corresponds to the expected theoretical loss. Indeed, assuming uncorrelated noise, the SNR of a sparse synthetic aperture system is proportional to the square root of the product of the number of transmit events N_t and the number of receive elements N_r [3].

$$SNR \propto \sqrt{N_t N_r} \tag{12}$$

(a)

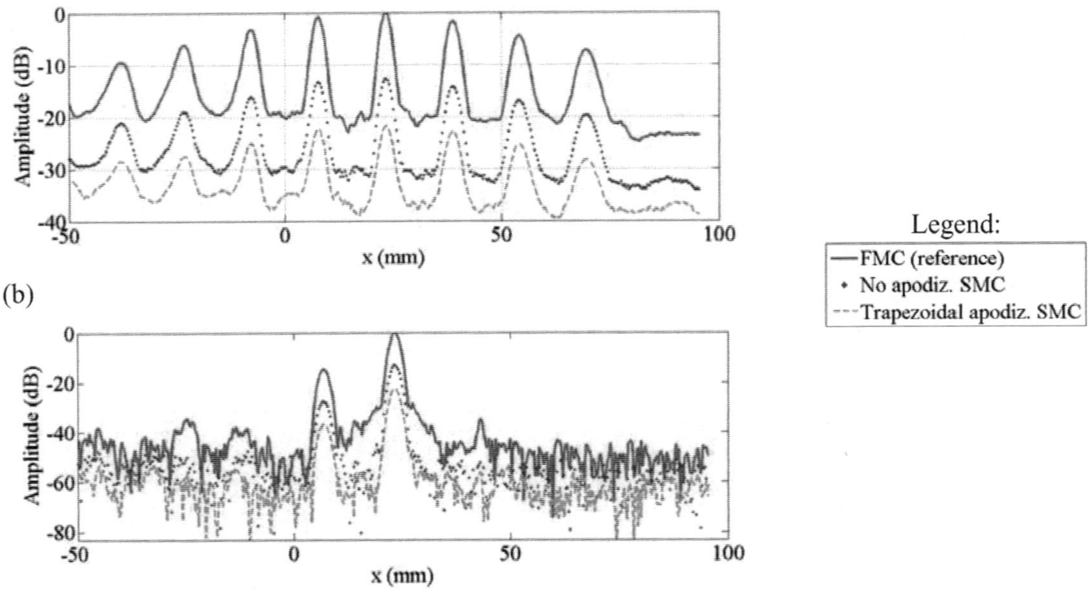

Figure 7: Quantitative measurement of TFM image quality. (a) Cumulative amplitude curves normalized by the maximum amplitude of the FMC-TFM image, (b) slice of image at z=50 mm depth, the amplitudes are normalized by the maximum amplitude of the FMC-TFM image.

Comparison of lateral resolution was performed on the images at -20 dB. In sparse-TFM images, the holes appear larger than on the FMC-TFM image. This increase amounts +6% and equals the increase of mainlobe width of the sparse array PSF measured above.

From these experimental results, it is clear that sparse-TFM image quality depends on several factors. Thus, the location of active elements must be chosen properly to avoid useless artifact and sidelobe noise. Applying weight to the elements in receive allows to lower significantly the number of firings but the main drawback is the associated loss in SNR. Indeed, the less combinations of transmit-receive elements used in the imaging algorithm, the less suppression of incoherent noise.

4. Improvement of the method and application to noisy samples

4.1. Concept of virtual sources

As seen in the previous section, the use of a single element per firing limits the acoustic power output. Furthermore, reducing the number of active transmit elements induces a loss of signal. Assuming random electronic noise, the loss of transmitted energy and signal results in a limited penetration depth and a loss of Signal to Noise Ratio (SNR). One method for increasing penetration depth and/or the SNR has been proposed by Karaman *et al* [9]. It consists in using multiple transmit elements for each transmit burst. The group of elements is de-focused by applying time delays. Thus, the elements simulate the radiation pattern of a single "virtual" source (VS) located above the array but with higher acoustic power (Figure 8(a)). Virtual sources can also be created under the array by applying focusing time delays to the subaperture (Figure 8(b)). The position of the VS with respect to the position of the array does not impact the detection of the flaw or the quality of the resulting TFM image if the position of the VS in the calculation of the focusing delays is taken into account.

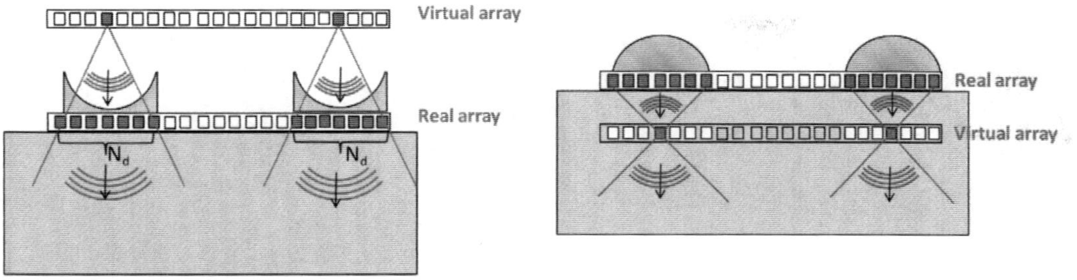

(a) (b)

Figure 8 : Principle scheme of virtual sources: (a) the virtual source is placed above the array, (b) the virtual source is placed under the array

The benefit of using "virtual" sources has been tested experimentally. First, the number of elements constituting a VS was determined by simulations performed with the CIVA platform. A perfectly spherical radiator should produce a uniform horizontal band in magnitude and constant horizontal phase bands [9]. Figure 9 provides the different simulated transmitted fields obtained when varying Nd. The fields are calculated along an arc of radius R = 50mm. As expected, when Nd = 1, the magnitude band (Figure 9(a)) is horizontal since a single element produces a cylindrical wavefront. Figure 9(b) and (c) show that for small values of N_d such as $N_d \leq 7$, the synthesized response of VS is very close to the one corresponding to a single element (Figure 9(a)). Larger subapertures ($N_d \geq 7$) produce bands that are no longer horizontal in a angular extent as large as for smaller subaperture (Figure 9(d), (e) and (f)). The response deviates significantly from a cylindrical wavefront, especially in range. Furthermore, edge wave diffraction appears and a loss of uniformity along the wavefront is measured. Thus, given the characteristics of the array probe and the results of the simulations, the optimal number of elements constituting a VS is 7.

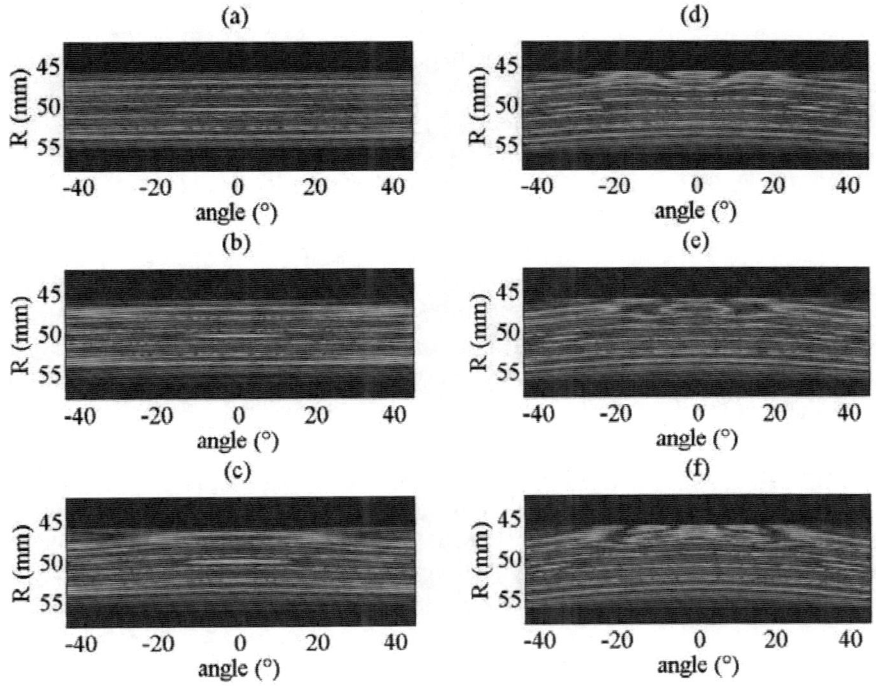

Figure 9: Simulated field transmitted by N_d defocused elements. (a) $N_d = 1$, (b) $N_d = 3$, (c) $N_d = 7$, (d) $N_d = 11$, (e) $N_d = 15$, (f) $N_d = 19$

The use of virtual sources increases the transmit power proportionally to the number of elements (N_d) constituting the de-focused subaperture. Karaman *et al.* demonstrated experimentally that if an isotropic radiation pattern is obtained from the "virtual" array, the amplitude of the resulting wave is increased in proportion of $(N_d)^{1/2}$ [9]. Measurements of the mean intensity of the simulated fields verify it. Figure 10 shows that the relation between the simulated transmit wave intensity and subaperture size can be approximated as $1.25\sqrt{N_d}$.

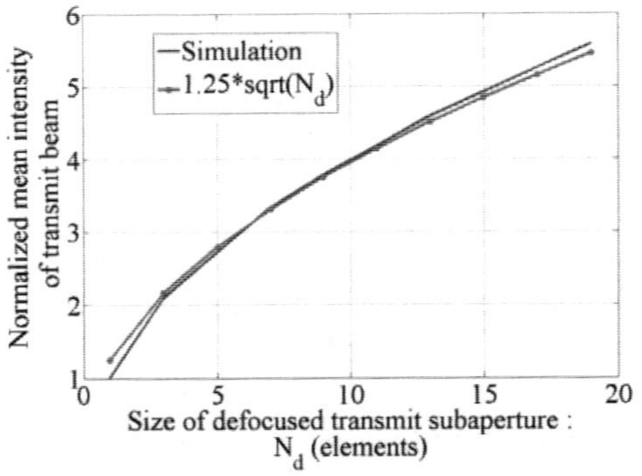

Figure 10 : Mean intensity of simulated defocused beam for different subaperture sizes

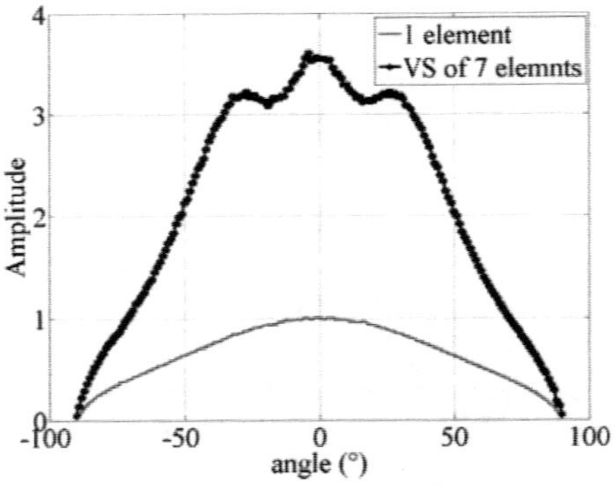

Figure 11: Simulated amplitude distribution for a single element (solid red graph) and for a virtual source of 7 defocused elements (dotted black graph)

Figure 11 represents, for a VS constituted of 7 defocused elements, the mean amplitude distribution measured on the simulated field. It allows anticipating the fact that employing VS can increase the transmitted energy by a factor 3.4 (≅ +11 dB) compared to the one delivered by a single element.

4.2. Experimental results

To demonstrate the benefit of using virtual sources, experimental testing were applied on a noisy sample. The same experimental set-up than previously was used on a cast steel sample. Figure 12(a) is a FMC-TFM image considered as the reference one. Figure 12(b) is a sparse-TFM image obtained after post-processing a SMC which sequences correspond to the ones provided by the optimization algorithm without apodization applied in reception. When comparing Figure 12(a) with Figure 12(b), it appears that the level of noise is higher in the SMC-TFM image than in the FMC-TFM image. As expected, the amplitude of the SDH is also lowered. A 13 dB loss can be measured on the echodynamic of the image (Figure 13, Table 3). Using a sparse matrix acquisition provided by a "virtual" array corrects partially this loss. Figure 12(c) was obtained by post-processing a SMC where active "virtual" elements are located at the same place that the real active elements used to get the sparse matrix acquisition of Figure 12(b). The image quality is clearly improved since the use of a "virtual" array allowed increasing the transmitted energy. Thus, slices of the resulting image almost fit the one of the FMC image (1 dB of difference). Employing virtual arrays allows obtaining images of quality similar to FMC-TFM images visually and quantitatively (Figure 13, Table 3) but with less firings.

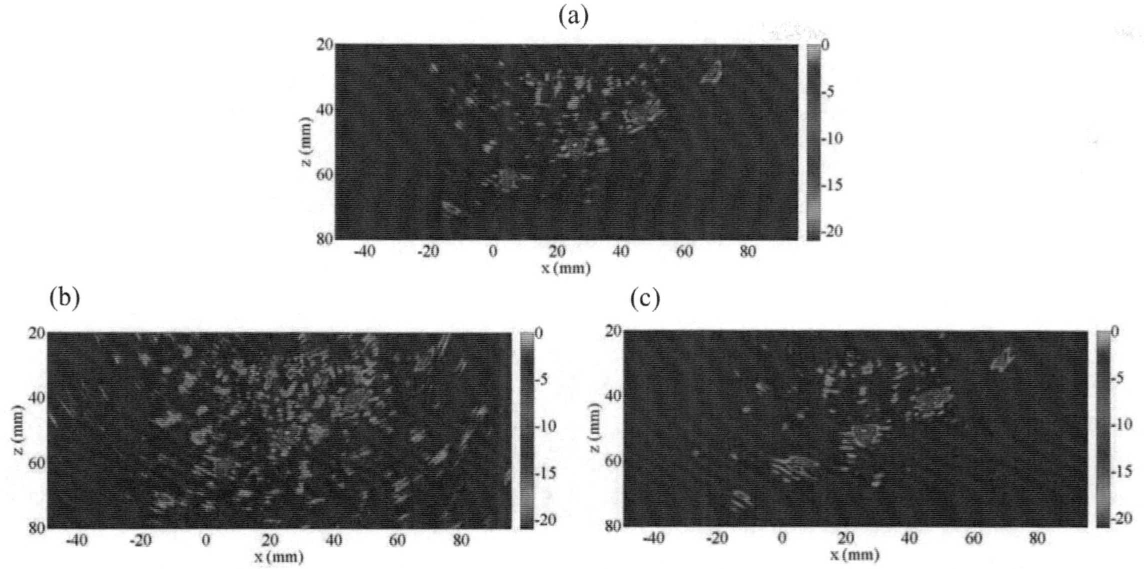

Figure 12 : Images of the cast steel sample obtained with (a) FMC-TFM, (b) real SMC-TFM, (c) virtual SMC-TFM

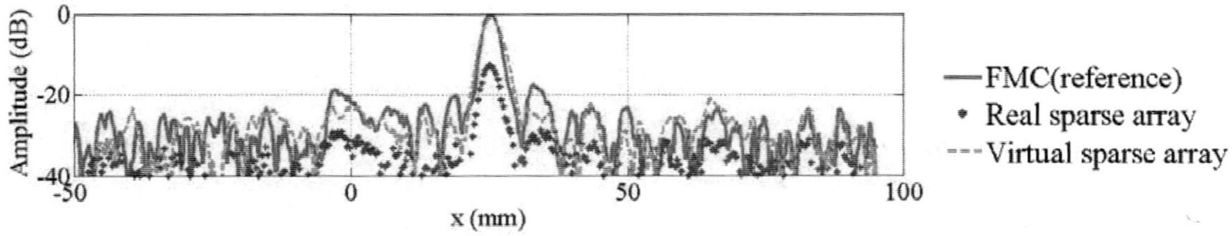

Figure 13: Slice of TFM images at depth z = 50 mm. The amplitudes are normalized by the maximum amplitude of the FMC-TFM image.

Table 3 : Quantitative characteristics of the cast steel sample TFM images

	FMC-TFM image	**Real Sparse-TFM image**	**Virtual Sparse-TFM image**
Number of firings	58	13	13
Amplitude of the central hole	0 dB	-13 dB	-1 dB

5. Conclusion

The two-way radiation pattern of an imaging system provides information about two main image quality parameters, lateral resolution and contrast. An algorithm based on optimization of the PSF has been proposed to the design of sparse array. The results provided by the algorithm and their experimental application show that the number of firings can be reduced by at least 4 without great loss of image quality compared to full array imaging. The PSF approach is based on far-field approximation. However, it has been tested in the near field of the array and applied to TFM imaging. Thus, although a loss of amplitude can be noticed, sparse-TFM images are found to be very close to TFM images obtained by post-processing FMC. Signal loss due to the reduction of active transmit elements is not a problem when imaging in non-noisy components contrary to materials presenting a high level of structural noise. To overcome this, virtual sources have been used. They consist in grouping adjacent elements and applying them time delay so they simulate the radiation pattern of a single element with higher acoustical power. This solution has shown its efficiency when imaging SDH in a cast steel sample. An increase of 12 dB has been noticed leading SMC-TFM image identical to full matrix TFM image.

The work presented in this paper suggests that TFM images of good quality can be obtained without processing the full array of data. Using sparse matrix capture allows to reduce the data volume by a factor 4. Thus, processing time is reduced and real-time application of TFM imaging can be considered.

6. References

[1] Orlakan O, Sanh Ergun A, Johnson J A, Karaman M, Demiciri U, Kaviani K Lee T H, Khuri-Yacub B T 2002 Capacitive micromachined ultrasonic transducers: next-generation arrays for acoustic imaging ? IEEE Transactions on Ultrasonics, Ferroelectrics, and Frequency Control, **49**, n°11 1596-1610

[2] Wilcox P D, Holmes C and Drinkwater B W 2006 Enhanced defect detection and characterisation by signal processing of Ultrasonic Array Data *Proc. 9th ECNDT*, Berlin

[3] Lockwood G R, Talman J R and Brunke S S 1998 Real-time 3-D ultrasound imaging using sparse synthetic aperture beamforming IEEE Transactions on Ultrasonics, Ferroelectrics, and Frequency Control, **45**, n°4 980-988

[4] Moreau L, Drinkwater B W and Wilcox P D 2009 Ultrasonic imaging algorithms with limited transmission cycles for rapid nondestructive evaluation IEEE Transactions on Ultrasonics, Ferroelectrics, and Frequency Control, **56** n°9 1932-1944

[5] Erstad J O and Holm S 1994 An approach to the design of sparse array systems, IEEE Symp. Ultrasonics, Proc. 1507-1510

[6] Lockwood G R, Pai-Chi Li, O'Donnell M and Foster F S 1996 Optimizing the radiation pattern of sparse periodic linear arrays IEEE Transactions on Ultrasonics, Ferroelectrics, and Frequency Control **43** 7-14.

[7] Nikolov M. and Behar V 2005 Analysis and optimization of medical ultrasound imaging using the effective aperture approach Cybern. Inform. Technol., **5**, n°2, 53-68

[8] Behar V and Adam D 2005 Optmization of sparse synthetic transmit aperture imaging with coded excitation and frequency division Ultrasonics **43** 777-788

[9] Karaman M, Pai-Chi Li and O'Donnell M 1995 Synthetic aperture imaging for small scale systems IEEE Transactions on Ultrasonics, Ferroelectrics, and Frequency Control **42** n°3, 429-442

[10] Chiao R Y and Thomas L J 1994 Analytic evaluation of sampled aperture ultrasonic imaging techniques for NDE IEEE Transactions on Ultrasonics, Ferroelectrics, and Frequency Control **41** n°4, 484-493

[11] Holmes C, Drinkwater B W and Wilcox P D 2005 Post-processing of the full matrix of ultrasonic transmit-receive array data for non-destructive evaluation NDT&E International, **38** 701-711

[12] Fidahoussen A, Calmon P, Lambert M, Paillard S and Chatillon S 2009 Imaging of defects in several complex configurations by simulation-helped processing of ultrasonic array data Review of Quantitative NonDestructive Evaluation **29** 847-854

[13] Bannouf S, Robert S, Casula O and Prada C 2010 Evaluation of multielement methods applied to complex geometries Review of Quantitative Non Destructive Evaluation **31**

[14] Frazier C H and O'Brien Jr W D 1998 Synthetic aperture technique with a virtual source element IEEE Transactions on Ultrasonics, Ferroelectrics, and Frequency Control, **45** 196-207

Towards semi-automated non-destructive evaluation

L Fradkin[1], V Zernov[1,6], G Elston[2], R Taneja[2], I Bell[3], D Lines[4], J Wharrie[4] and P Fitzgerald[5]

[1] Sound Mathematics Ltd., 11 Mulberry Close, Cambridge, CB4 2AS, UK
[2] Doosan Power Systems, Porterfield Road, Renfrew, PA4 8DJ, UK
[3] National Instruments , Measurement House, Newbury Business Park, London Road, Newbury, Berkshire RG14 2PZ, UK
[4] Diagnostic Sonar Ltd., Baird Road, Kirkton Campus, Livingston, EH54 7BX, UK
[5] LabVIEW Certified Developer, Lewkenors Mill, Denham, Bury St Edmunds, Suffolk IP29 5EF, UK
[6]Now at SmartOdds, 53-79 Highgate Road, London, NW5 1TL, UK

E-mail: l.fradkin@soundmathematics.com

Abstract. A demonstrator has been developed showing feasibility of semi-automatic characterisation of large planar flaws in steel using ultrasonic transducer arrays. The unit is based on a real-time ultrasonic imager deploying National Instruments hardware and software, is connected to an IMASONIC linear phased array containing 128 elements and incorporates a novel flaw characterisation algorithm, which is a model-based variant of Total Focusing Method, taking into account undulations in inspection surface. It has been shown to process RF data collected in immersion reasonably fast and be capable of detecting and characterising with reasonable accuracy large planar defects.

1. Introduction

We have addressed a high level, long-term challenge of the ultimate deployment of an innovative ultrasonic unit for flaw characterisation and interpretation that would lead to cost effective and reliable semi-automatic NDE (Non-Destructive Evaluation). No existing instrumentation offers real-time automatic flaw diagnostics and at present, while some phased array ultrasonic inspections themselves are automated the data collected are still being interpreted by human inspectors, leading to variability in reporting [1]. The most surprising outcome of the above study conducted by TWI (The Welding Institute) is that inspectors experience the greatest difficulty when characterising large planar cracks. Probably less surprisingly the most difficult cracks to identify are those normal to inspection surface.

While in order to carry out crack characterisation inspectors rely mostly on TOFD (Time of Flight Diffraction) approaches pursued by those who work towards automating crack characterisation can be broadly divided into pure signal processing procedures, such as those based on CS (Compressed Sensing) [2] or FMC (the Full Matrix Capture) [3], [4] which are best suited for dealing with specular reflections and more general but also more time-consuming model-based data processing algorithms, such as those desciribed in [5] and [6].

Content from this work may be used under the terms of the Creative Commons Attribution 3.0 licence. Any further distribution of this work must maintain attribution to the author(s) and the title of the work, journal citation and DOI.
Published under licence by IOP Publishing Ltd

In this study we work with real data collected in immersion, with the main source of error being undulations in the inspecion surface and show that a simple modification of a TFM (Total Focusing Method) supported by a simple model of a large planar crack can produce a reasonably fast algorithm for crack characterisation. The paper is organised as follows: in the next section we describe the demonstrator developed, experiments conducted and signals collected, we then report our findings and compare our images to the ones obtained using the standard FMC (Full Matrix Capture) algorithm as implemented in our demonstrator. We finish by discussing our results and future plans.

2. Description of the demonstrator and experiments

History and issues surrounding real-time ultrasonic array imaging using FMC and TFM have been described in many publications, see e.g. [7]. The demonstrator built during this study took full advantage of flexible and scalable acquisition hardware modules, thus reducing instrument cost and size. The acquisition time has been reduced by introducing significant levels of parallelism, using standardised PC-based modular hardware. Generic hardware (the FlexRIO family from National Instruments) based on FPGAs (Field Programmable Gate Arrays) has been combined with customized analogue hardware (NI-5752) to produce a 32 channel 12-bit 50 MSPS acquisition module with digitally controlled swept gain. The multiplexer used on transmit has been implemented as a stackable PCB module to multiplex to a 128 element IMASONIC linear transducer array. The module is able to multiplex the 32 receive channels over all 128 elements. FPGAs handle the high data rates from many parallel channels. While existing techniques take two or more seconds to process each image the new architecture achieves acquisition at 20Hz and higher. The IMASONIC array employed in the study has had the elementary pitch of 0.8 mm, inter element space of 0.25 mm and total active length of 102.15 mm. All elements have been employed to transmit simultaneously the signal with the centre frequency of 2 MHz $\pm$ 10%, the bandwidth $\geq$ 55 % and pulse duration < 1500 ns (see figure 1).

The specimen used in experiments was a steel block, 30 mm thick, 200 mm wide, 350 mm long, containing four surface-breaking notches and four embedded ones, some tilted some – non-tilted. The full description of the notches is given in table 1. The experiments have been performed in immersion, with the water temperature of 22.0^{0}C and water path standoff distance of about 13 mm. The geometry of the experiment is presented in figure 2.

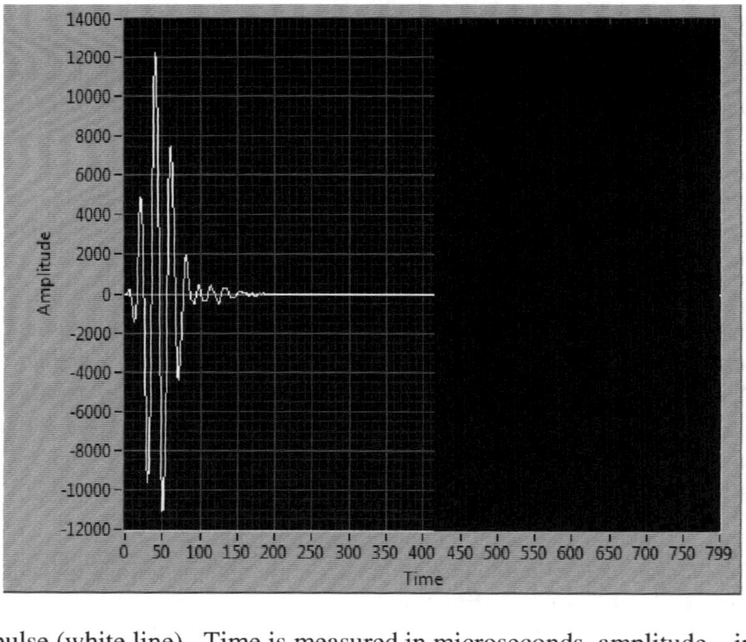

Figure 1. Input pulse (white line). Time is measured in microseconds, amplitude – in arbitrary units.

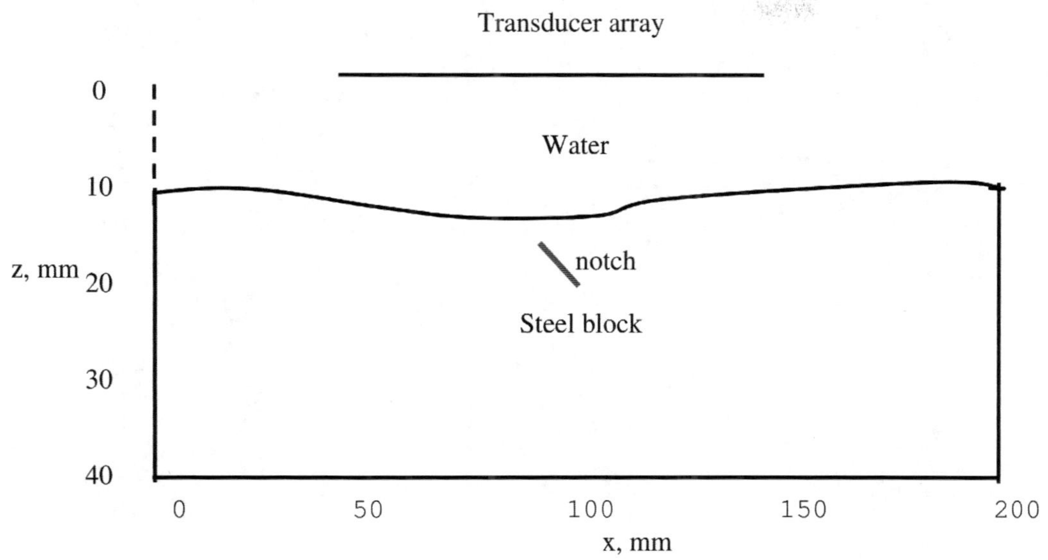

Figure 2. Schematic of the experiments

The present version of the demonstrator generates data in the form of .png files. These are used to be processed by either standard or modified TFM and stored for future use. A LabVIEW application has been written that can read these .png files and convert them into a Full Matrix [F_{ijk}, i, j = 1,2,...128, k = 1,..., Alen] of A-scans, where the first index denotes the transmitter, the second – the receiver and the third – the time sample. Alen has been normally chosen to be 800 to deal with pulse trains 800 μsec long. The standard FM produces images of specimen sections utilising the intensity function

$$I(\mathbf{x}) = \left| \sum_{tx,rx} h_{tx,rx}(t_{tx,\mathbf{x},rx}) \right| \tag{1}$$

where $h_{tx,rx}(t)$ is the Hilbert transform (analytic representation) of the signal $A_{tx,rx}(t)$ and $t_{tx,\mathbf{x},rx}$ is the time of travel from a transmitter tx to receiver rx through a node $\mathbf{x}$ =(x,z). The nodes are evenly spaced throughout each section. The Hilbert transform is defined by

$$H(A)(t) = \frac{1}{\pi} p.v. \int \frac{u(\tau)}{t-\tau} d\tau$$

provided the integral exists as a principal value. The summation in (1) is carried out for each possible transmitter–receiver pair in the array. Even when the input signal is a perfect delta function rather than the pulse presented in figure 1 the intensity function can produce misleading results: In the worst case scenario, if the transducer array is in direct contact with the specimen, only one transducer element is used in both transmit and receive and there is only one point scatterer, all specimen points lying on an arc of a circular wave front radiated by this element and reaching the scatterer are characterised by the same time of arrival and therefore are assigned the same intensity. Inspecting in immersion and involving more than one element reduces this possibility but does not eliminate it. The fact that different portions of a pulse arrive at different moments in time increases the probability of false indications.

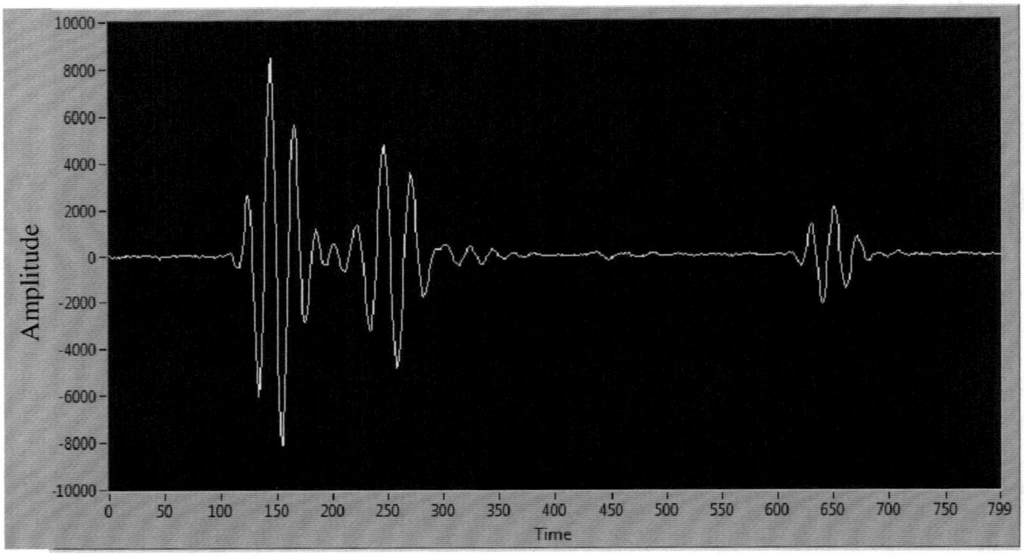

Figure 3. A typical pulse train recorded at array element 17 (elements are numbered left to right). The first pulse – echo from the upper surface of the specimen of the pulse transmitted by element 17, the second pulse – surface echo of the pulse transmitted by element 1, the third pulse – backwall echo of the pulse transmitted by element 17. The backwall echo of the pulse transmitted by element 1 arrives later or not at all. Time is measured in microseconds.

3. Semi-automatic crack characterisation

As mentioned at the end of the previous section, applying the intensity function (1) different points on the grid that surrounds a scattering point can pick up different portions of the same scattered pulse. This can lead to a smudged image of the scatterer and maybe, to an error in its positioning. The brightest point in the image can be close to, but not necessarily on, the scatterer. For this reason, Wiener deconvolution has been tried to identify an impulse response function but it has produced no discernibly superior results and is not employed in the final algorithm. However the above considerations have been combined with a priori knowledge that the notches are linear features a few wave lengths in extent and thus, when in TOFD configurations, can produce two bright spots around edge diffraction points. The resulting LabVIEW application first combines image and signal processing algorithms to implement a model-based modification of TFM to identify such spots. It then joins their centres by a straight segment to characterise the defect in terms of its orientation (in degrees) with respect to the inspection surface, extent (in mm) and depth (in mm), that is, the shortest distance from a defect end to a specimen's surface (the inspection surface or else the backwall). It has been found that the diffraction points are best identified by utilising the intensity function (1) with the summation carried over a portion of the array, which we call a fictitious aperture, rather than the whole array. This is due to the fact that when the full aperture is used the specular reflections from the inspection surface can be very intense and mask diffraction signals from defect edges.

Journal of Physics: Conference Series **457** (2013) 012008 doi:10.1088/1742-6596/457/1/012008

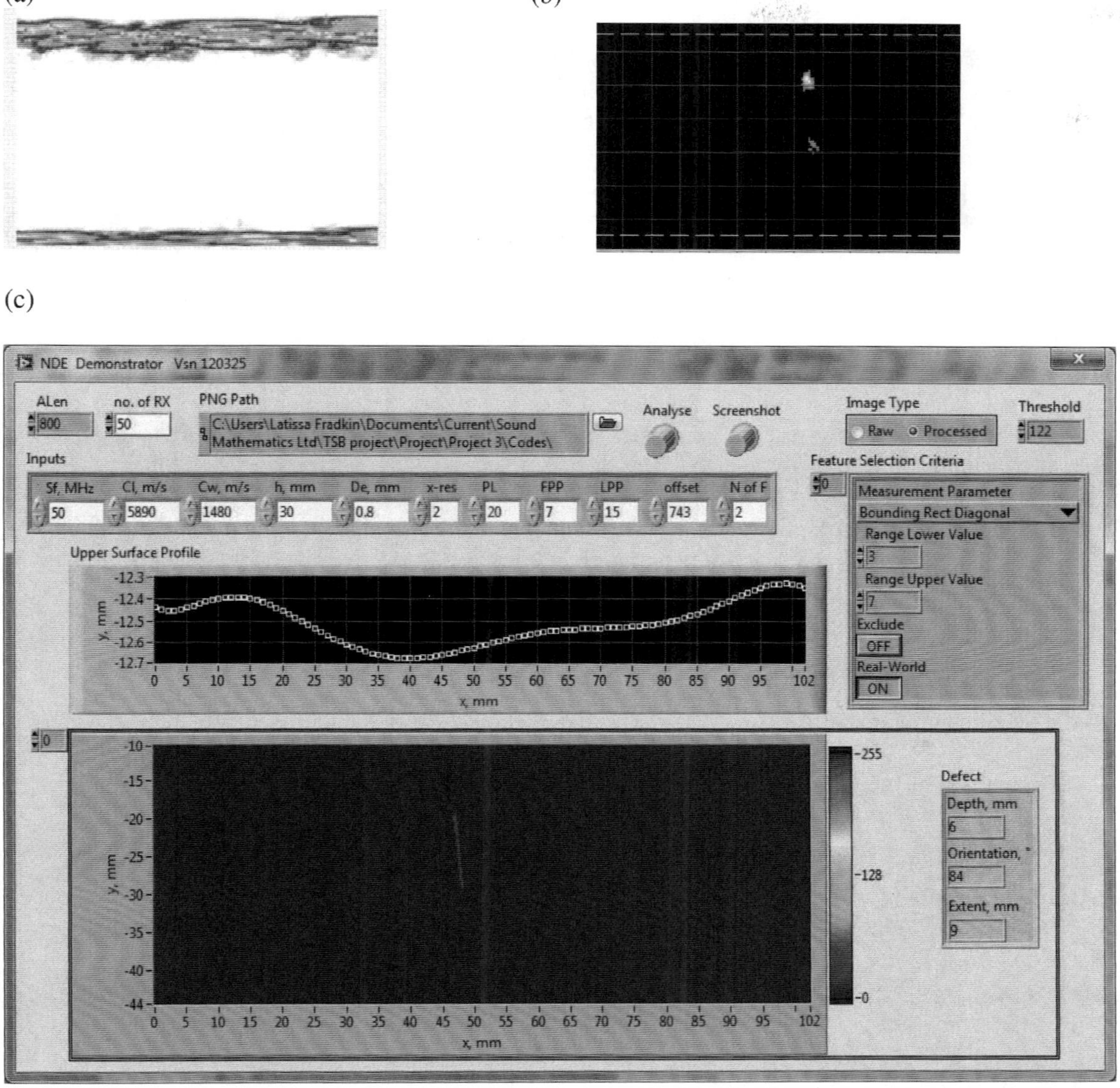

Figure 4. Comparison of images of a non-tilted embedded defect, imaged from the notched block side (notch 3 in table 1). (a) - the image obtained with a standard TFM algorithm, (b) – the image obtained with a model-based version of TFM, (c) - the resulting report. (a) demonstrates that in this case the straightforward application of TFM produces no defect image. (b) presents in black and white two diffraction spots identified using the model-based version of TFM. The report in (c) contains the profile of the specimen's upper surface and the defect image suggested by the diffraction spots as well as the associated defect parameters (bottom right) and parameters of the image processing filters that have been used to generate (b) (top right).

A typical pulse train is presented in figure 3 and a typical report - in figure 4. Typical results are summarised in table 1. Extensive data analysis has shown that to achieve successful defect characterisation several aspects of the problem require attention:

1. The defect images are sensitive to undulations in surface profile. Although the undulations are small (about 0.5 mm change over 100 mm length), due to the large difference in ultrasonic

wave speeds in water and steel, they lead to significant change of the ray paths – see figure 5, where different approximations of the upper surface are used, a polynomial and a constant, respectively. Numerical experimentation has been used to establish the range of interpolation parameters within which the sensitivity of defect images to these parameters is at its lowest. In future interpolation will be effected using cubic splines.

2. As table 1 illustrates, there are issues with characterising surface-breaking defects when imaging form the notched surface. Analysis has shown that these are due to failing to detect the diffraction points lying in the inspection surface. Similar difficulties face human inspectors. However, our results suggest that identification can be improved by employing image processing filters which incorporate *a priori* knowledge of properties of scattered signals.

3. Occasionally we see false indications. Therefore all solutions suggested by our application should be presented to a human operator to allow him/her to select those that appear most reasonable.

4. When analysing data pertaining to all studied defects it has been found that the most advantageous length of fictitious aperture is 20 mm. In each case position of this aperture is selected automatically.

Table 1. Comparison of estimated and actual characteristics of notches imaged from either flat or notched side of the test steel block.

Notch no/type/ inspection surface	Estimated/specified extent	Estimated/specified orientation	Estimated/specified Depth
1/embedded/flat	9mm/10 mm	$70^0/70^0$	7mm/5 mm
1/embedded/notched	10 mm/10 mm	$70^0/70^0$	9 mm/5 mm
2/embedded/flat	5 mm /5 mm	$70^0/70^0$	5 mm /5 mm
2/embedded/notched	4 mm/5 mm	$70^0/70^0$	8 mm/5 mm
3/embedded/flat	10mm/10 mm	$90^0/90^0$	3 mm/5 mm
3/embedded/notched	9 mm /10 mm	$84^0/90^0$	6 mm/5 mm
4 /embedded/flat	4.5/5 mm	$90^0/90^0$	4.5 mm/5 mm
4 /embedded/notched	4 mm/5 mm	$90^0/90^0$	7 mm/5 mm
1/surface-breaking/flat	4 mm/5 mm	$60^0/90^0$	0 mm/0 mm
1/surface-breaking/notched	5 mm/5 mm	$90^0/90^0$	1 mm /0 mm
2 /surface-breaking/flat	8 mm/10 mm	$90^0/90^0$	1 mm /0 mm
2/surface-breaking/notched	5 mm /10 mm	$90^0/90^0$	5 mm /0 mm
3/surface-breaking/ flat	5 mm/5 mm	$80^0/70^0$	0 mm/0 mm
3/surface-breaking/notched	5 mm/5 mm	$68^0/70^0$	2 mm/0 mm
4 /surface-breaking/flat	10 mm/10 mm	$70^0/70^0$	0 mm /0 mm
4/surface-breaking/notched	9 mm/10 mm	$64^0/70^0$	2 mm/0 mm

(a)

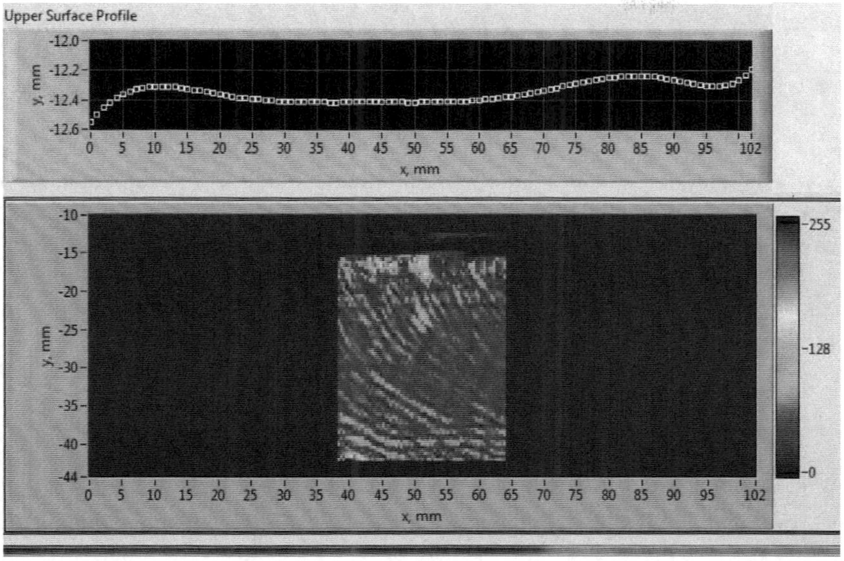

(b)

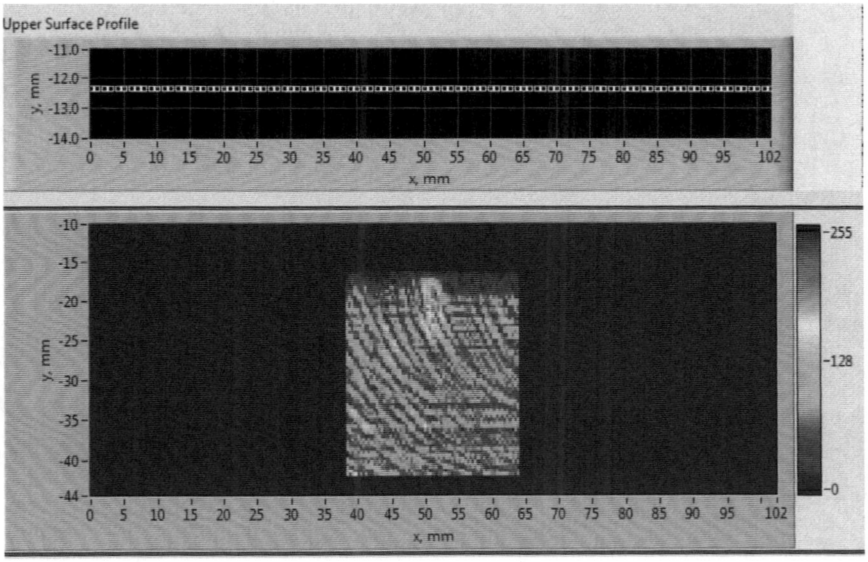

Figure 5. (a) Upper surface profile approximated by a 10th degree Polynomial and resulting raw image. (b) Upper surface profile approximated by a 0th degree polynomial (constant) and resulting raw image.

4. Speed of execution

The time required to obtain one 100 mm x 10 mm image frame using 30000 pixels using the proposed algorithm as implemented in MATLAB takes about 10 s. This compares favourably with the speed of TFM codes developed by Bristol group [5]. At present the algorithm is embedded in the LabView code in the form of a MathScript. Surface profiling does not depend on the spatial resolution, and for this module MathScript shows better performance. However, for the ray tracing and imaging, the performance of LabVIEW implementation is up to 10 times slower that of the MATLAB version. On the other hand, the speed of the LabView implementation of the standard TFM running on our demonstrator matches its speed of data acquisition at 20 Hz or higher. Thus, a significant speed-up of our application could be achieved by re-writing MathScript modules in LabVIEW. An alternative fast version could be written in C.

5. Conclusions

A model-based version of Total Focusing Method has been developed and implemented in LabVIEW for semi-automated characterisation of large isolated planar cracks in stainless steel. The reported feasibility study has shown that simulating configurations which produce diffraction from defect edges usually leads to reliable defect sizing, even though further work is required to study and reduce probability of false indications. The capability of application was demonstrated using a steel block with realistically undulated surfaces and containing four embedded and four surface-breaking relatively large planar notches, some tilted and some non-tilted. It is possible to extend the procedure to other types of defects and geometrical configurations, developing a comprehensive library of generic models for eventual deployment in a portable probe capable of acting as a real-time assistant to an ultrasonic inspector and interpreter.

The proposed solution for semi-automatic characterisation of safety critical defects, which would result in clear and unambiguous reports, could support the existing fleet of nuclear reactors as well as the new build. It could be later spun-out into other industries, such as rail or oil and gas.

Acknowledgments

This project was funded by Technology Strategy Board under their Nuclear R&D Feasibility Studies Program and the authors are indebted to the Board for their imaginative strategies for setting up creative consortia. The authors are also grateful to the TSB monitoring officer Tom Harris for his critical feedback and constant support.

References

[1] Schneider C and Bird C 2009 *Proc 4th European-American Workshop on Reliability of NDE*, 24th – 26th June 2009 Berlin Germany
[2] Champagnat F, Goussard Y and Idier J 1996 *IEEE Transactions on Signal Processing* **44**(12) 2988
[3] Holmes C, Drinkwater B W and Wilcox P D 2008 *Ultrasonics* **48** 636
[4] Hunter A J, Drinkwater B W and Wilcox P D 2008 *IEEE Trans Ultrason Ferroelectr Freq Control.* 2008 **55**(11) 2450
[5] Hunter A J, Drinkwater B W and Wilcox P D 2011 *IEEE Trans Ultrason Ferroelectr Freq Control* **58**(2) 414
[6] Marengo E A, Fred K G and Simonetti F 2007 *IEEE Transactions On Image Processing* **16**(8) 1967
[7] Lines D I A, Wharrie J and Hottenroth J 2010 *Proc.ASNT Fall Conference*, Houston, 15-19 November 2010.

Journal of Physics: Conference Series **457** (2013) 012009 doi:10.1088/1742-6596/457/1/012009

Emergence of the coherent reflected field for a single realisation of spherical scatterer locations in a solid matrix

V J Pinfield and R E Challis

Electrical Systems and Optics Division, Faculty of Engineering, University of Nottingham, University Park, Nottingham NG7 2RD, UK

Email: valerie.pinfield@nottingham.ac.uk

Abstract. The acoustic field reflected from a region containing spherical scatterers is most often estimated by use of the coherent field; that is, the field resulting from the summed scattered fields from the scatterers, averaged over all possible configurations of scatterer locations. It is this ensemble-averaged coherent field which is equivalent to the field reflected from a homogeneous medium with properties which can be derived mathematically using ensemble-averaging techniques. Such properties include the effective density, and the effective wavenumber, which can be derived from multiple scattering theories, or by other homogenisation methods. Experimentally, although ensemble-averaging can be effected in practice in fluid systems due to the motion of the scatterers during the measurement time-scale, measurements in solid materials have fixed locations of scatterers. Averaging can only be achieved by using "large" sample areas, multiple samples or measurements in different locations, or "large" receiver areas. However, in the context of NDE applications we are interested in the field resulting from a specific region of material, rather than the average over a large region. Our study addresses the question of when the coherent field (resulting from averaging over many scatterer configurations) can be used as an accurate description of the field reflected by a region of scatterers at fixed locations. In this paper we present results of simulations of the scattered reflected field from a region of solid material containing spherical cavities. Simulations of single realisations of scatterer locations are compared with the coherent field, to demonstrate the validity or otherwise of the use of the coherent field to describe the response of a particular configuration of scatterers.

1. Introduction

The field of acoustic wave propagation in heterogeneous materials has a long history and remains a topic of study due to its range of significant applications. One such application is the use of ultrasonic non-destructive testing techniques for aerospace components, which has prompted the current work. Our interest is in the signal backscattered from a region of small cavities in a composite material.

The majority of published work concerned with acoustic propagation in heterogeneous media considers the coherent field, that is the field which has the same characteristics as that propagating in a homogeneous medium, but with modified properties. This coherent field emerges only on averaging the locations of scatterers, a process known as ensemble-averaging, or homogenization [1]-[5]. The properties of the equivalent homogeneous medium or effective medium resulting from the coherence condition can be used to define the response of the material. These properties include the effective wavenumber, which is widely used to obtain the speed of wave propagation in materials. In an experimental measurement, the ensemble averaging process may be effected by measurements over a

Content from this work may be used under the terms of the Creative Commons Attribution 3.0 licence. Any further distribution of this work must maintain attribution to the author(s) and the title of the work, journal citation and DOI.
Published under licence by IOP Publishing Ltd

large area of sample, multiple samples, a large received area, or by a time-averaging (if scatterers are mobile). In some circumstances however, we are interested in the response of a particular region of material, which may not be coherent, since the scatterers are fixed and the region small.

In this study, we investigate the conditions for coherence for a particular configuration, in which the response of a plane slab of scatterers in a solid matrix is simulated, using a single receiving point. Thus we eliminate the effects of averaging at the receiver, but consider the response of a planar region. A range of simulations has been carried out to investigate the emergence of the coherent field over a range of cavity radii and concentrations (number density). We establish the dependence of the degree of incoherence of the field with these parameters compared with the wavelength. We also investigate the variability in the degree of incoherence of the signal.

2. The models

The computational models for the study of the backscattered field have been presented in previous papers [6]-[7] and only a brief summary is presented here.

2.1. Discrete scatterer model

Scatterers are located randomly within a solid matrix, in a slab region as shown in Figure 1a. A transmitting and receiving surface is coupled directly to the solid and its response is assumed to relate to the normal displacement at the surface. The incident field at each scatterer is assumed to be the plane wave transmitted from the surface. The discrete scatterer model simulates the field at a point on the receiving surface obtained from the sum over the scattered fields from each scatterer. The scatterers are taken to be spherical cavities of identical radius. Only single scattering is considered, and the anisotropy of the scattered field is included using the far-field scattering amplitude, derived from Rayleigh partial wave expansions according to the method of Ying and Truell [8].

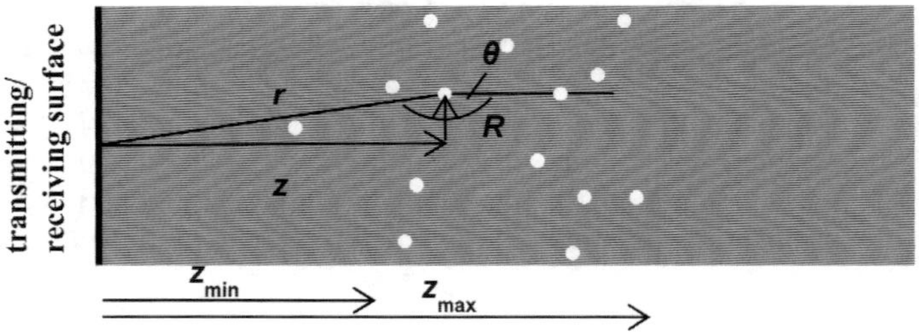

(a)

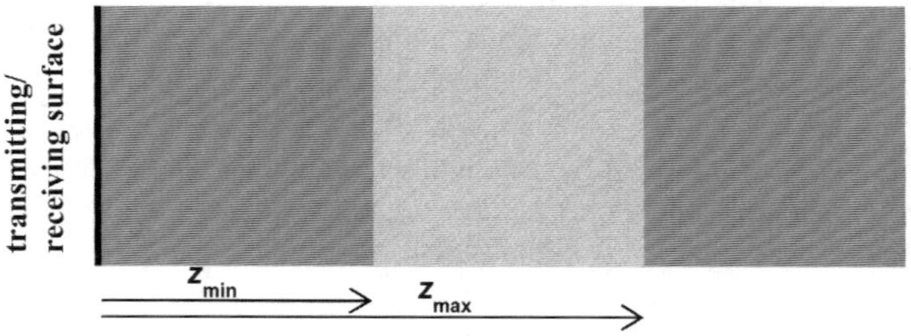

(b)

Figure 1 Configuration of the models: (a) discrete scatterer model (above dashed line) and (b) effective medium model (below dashed line). The coordinates (z,R) of a cavity relative to the receiving point are shown.

The normal displacement at the receiving surface is given by

$$\mathbf{u}_{z,\mathrm{mult}} = \mathbf{u}_{z,\mathrm{inc}} \sum_{j=1}^{N_{sc}} e^{ikz_j} \cdot \frac{e^{ikr_j}}{r_j^2} \left[-\left.\frac{\partial f(\theta)}{\partial \theta}\right|_{\theta_j} \sin\theta_j - ikz_j \cdot f(\theta_j) \left\{ 1 - \frac{1}{ikr_j} \right\} \right] \tag{1}$$

where the far-field amplitude is related to the scattering coefficients by

$$f(\theta) = \frac{1}{ik} \sum_{n=0}^{\infty} (2n+1) A_n P_n (\cos\theta) \tag{2}$$

The wavenumber in the solid matrix is denoted by k and the angle, θ, is taken relative to the forward z-direction. P_n are the Legendre polynomials and r is the distance from the scatterer to the receiving point. The normal displacement of the transmitted plane wave at the transmitting surface is $\mathbf{u}_{z,\mathrm{inc}}$. The notation for the scattering coefficients, A_n, is adopted from earlier references, and is presented in [9].

In the discrete scatterer model, there are no interfaces between the solid matrix and the region containing scatterers. The model predicts the response from a single realisation of scatterer locations for a particular cavity radius and number density (expressed as a volume fraction).

2.2. Ensemble-average model

An estimate of the ensemble-average limit, that is the coherent field, is obtained by numerical integration over the discrete scatterer model field equations using random locations for scatterers.

2.3. Effective medium model

In the effective medium model, the region containing scatterers is replaced by a homogeneous material (Figure 1b) whose properties are assigned using ensemble-averaged scattering theory formulations. Thus in this model, there are interfaces at the front and back of the slab region at which the material properties change. The effective density used in the model is the volume averaged density.

$$\rho_{\mathrm{eff}} = \rho(1-\phi) \tag{3}$$

where ϕ is the volume fraction of cavities and ρ is the density of the solid matrix. The Foldy wavenumber [1] was used to obtain the wave speed, taking the same partial Rayleigh wave analysis for the scattering coefficients [9].

$$\left(\frac{K_{\mathrm{eff}}^2}{k^2} \right) = 1 + \frac{3\phi}{k^2 a^3} f(0) \tag{4}$$

where a is the radius of the cavities and K_{eff} is the effective wavenumber of the region. These properties are used to define an effective impedance for the layer, permitting standard reflection and transmission coefficients to be assigned to the interfaces. Thus the reflected coherent signal can be expressed in the usual way as

$$\mathbf{u}_{z,\mathrm{eff}}(\omega) = \mathbf{u}_{z,\mathrm{inc}}\, e^{2ikz_{\mathrm{min}}} \cdot r_{12} \left[1 - t_{12} t_{21} e^{2ikd} \left\{ 1 - r_{12}^2\, e^{2ikd} \right\}^{-1} \right] \tag{5}$$

where $d = z_{\mathrm{max}} - z_{\mathrm{min}}$ is the thickness of the layer and r_{ij}, t_{ij} are the displacement reflection and transmission coefficients between regions i and j. Region 1 is between the transmitter and the slab, and region 2 is the slab region whose effective properties are under consideration.

These models incorporate some simplifications in order to enable the simulation of a large slab region. Other workers have studied the coherent field using techniques such as finite difference methods, which are constrained to smaller scale simulations, but are able to adopt higher order assumptions on the incident field for example. Since we are interested in the emergence of the coherent field and its dependence on the length scales of the system, we consider the field at a single point on the receiving surface in order to remove any effects of averaging at the receiver.

2.4. Numerical calculations

Numerical calculations were carried out using Matlab. The fields were calculated in the frequency domain with a sampling frequency of 50 MHz and 1024 samples. Time-domain results were obtained using experimentally sampled transmit-receive signals to model the transmitted signal for a 5,10 MHz centre-frequency transducer. A range of cavity radii from 5-20 μm and concentrations by volume of 1-20% were simulated. Each discrete scatterer model simulation represents a single realization of cavity locations, all of the same size. The solid matrix was assigned properties appropriate to an averaged fibre-resin composite, with a longitudinal wave speed of 3035 ms^{-1}, a density of 1564 kgm^{-3}, and a shear modulus of 3.6 GPa. The simulation aims to represent the response of a slab, but must be constrained in the lateral direction for calculation purposes. Tests showed that a maximum radial coordinate of R_{max}=20 mm (see Figure 1) was sufficient to capture the full response. Other system parameters are listed in Table 1.

Table 1. System parameters.

Distance z_{min} /mm	2
Layer thickness / mm	1
Cavity conc / v/v%	1, 2, 5,10, 20
Cavity radius / μm	5.0, 7.9, 10.0, 15.9, 20.0
Millions of cavities at 20 v/v% (respectively with radius)	480, 120, 60, 15, 7.5

All time-domain results have been plotted time-shifted to the arrival of the first reflection from the front of the layer and have been scaled by concentration for ease of comparison. Frequency-domain results were windowed in the time domain to smooth out the edge-wave ripples caused by truncation of the region. Fuller details of the simulation are given in [6]-[7].

3. Results

3.1. Time-domain results

Results of the models are shown in Figure 2 in the time-domain using signals with centre frequencies (c.f.) of 5 and10 MHz. The results have been scaled by concentration for convenience of comparison. In each figure, the coherent response (from the effective medium model) can be seen to display reflections from the front and back of the layer with the appropriate time delay between them. However, the discrete scatterer responses (in black) show varying degrees of agreement with this coherent field. In Figure 2a, the discrete scatterer response for 10 μm radius cavities at 2%, with a 10 MHz c.f. signal, shows a relatively high degree of incoherence, although the initial response is similar to the front face reflection in the coherent field. There is a significant signal between the two coherent reflections, and after them, and the back-wall reflection in the coherent response is not clearly identifiable. At a higher concentration Figure 2b, some of the incoherence is reduced, especially the signal received after the termination of the coherent field. Reducing the radius has a more marked effect at improving coherence. Comparison of Figure 2b (for 10 μm radius) and Figure 2c (for 5 μm radius) shows that the reduction in radius produces a nearly-coherent field at 10% concentration (by volume) with a 10 MHz c.f. signal. It should be noted that since these are at the same *volume* concentration, the number density is significantly larger at the smaller radius. There is, however, a difference in arrival time of the back wall echo between the discrete scatterer and effective medium models, because the discrete scatterer model is a single scattering model, and is not able to predict the change in effective wave speed due to the porosity, which does occur in the coherent field. A reduction in the centre frequency of the signal also has a significant effect in the emergence of the coherent response from the discrete scatterer model, as seen by comparing Figure 2b (a 10 MHz c.f. signal) and Figure 2d, which shows the response with a 5 MHz c.f. In this case, both front and back-face reflections of the coherent field are produced from this single realization of scatterers in the discrete scatterer model. The difference in arrival time is also clear here, but the plane-wave-like behavior arising from the discrete scatterer model for these single realisations is apparent. Comparisons between various pairs of these figures indicate that coherent field emerges at lower frequency, smaller cavity radius (at constant volume concentration) and higher concentrations.

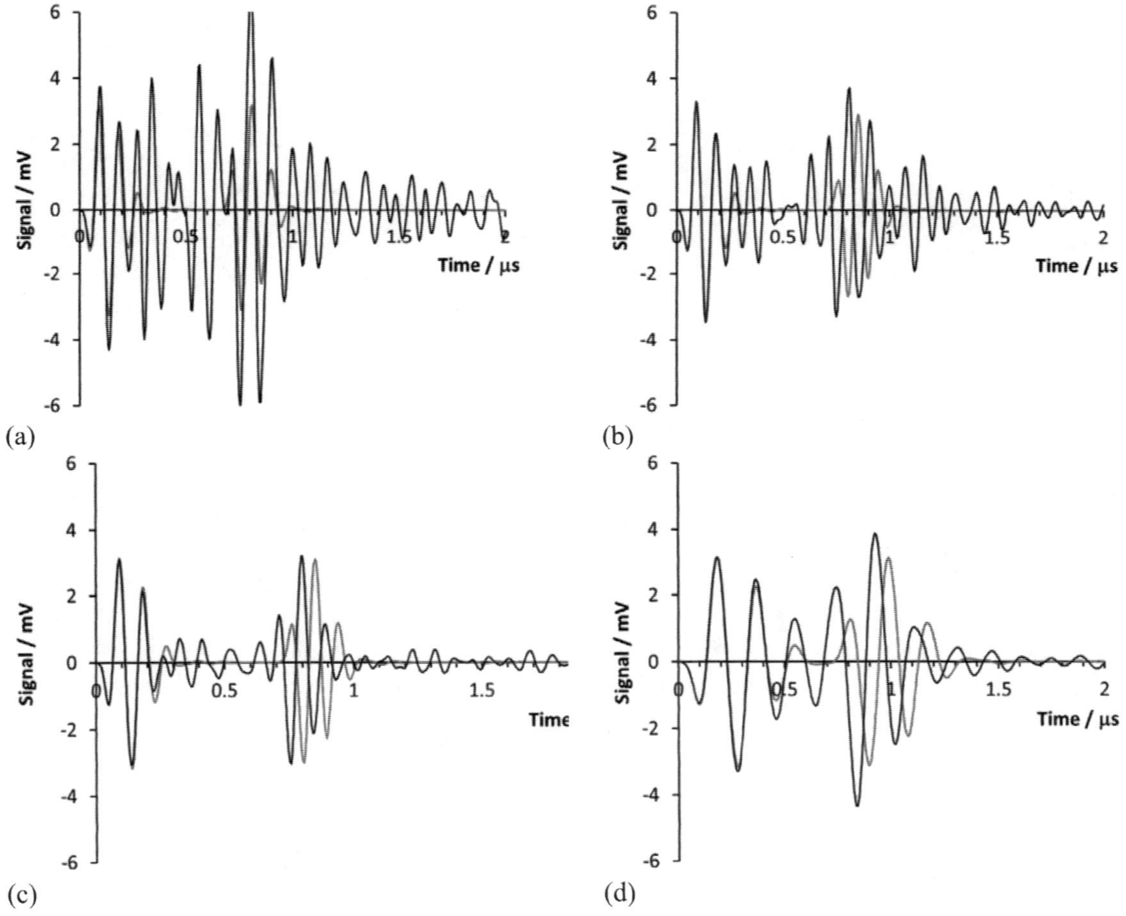

Figure 2 Time-domain results of discrete scatterer (black) and effective medium model (red) simulations for various cavity radii, concentrations and signal centre frequencies (c.f.) (a) 10 μm radius, 2 v/v% with 10 MHz c.f. (b) 10 μm radius, 10 v/v% with 10 MHz c.f. (c) 5 μm radius, 10 v/v% with 10 MHz c.f (d) 10 μm radius, 10 v/v% with 5 MHz c.f.

3.2. Frequency-domain results

In the frequency domain, the response to a time-domain impulse is typical of that of a thick layer for the effective medium model, showing regularly spaced resonance peaks (Figure 3). The integrated ensemble-average response is also shown (in blue) illustrating a close agreement with the effective medium model, with a slight difference in peak spacing due to the difference in sound speed between the ensemble average (single scattering model) and the effective medium model. Figure 3 also shows the response from the discrete scatterer model (in black) for a single realisation of scatterer locations, for a cavity radius of 10 μm at a concentration of 2% (scaled by concentration). Whilst the discrete scatterer response matches that of the effective medium model at the lowest frequencies, i.e. is a coherent response, it rapidly diverges as the frequency increases, showing increased incoherence.

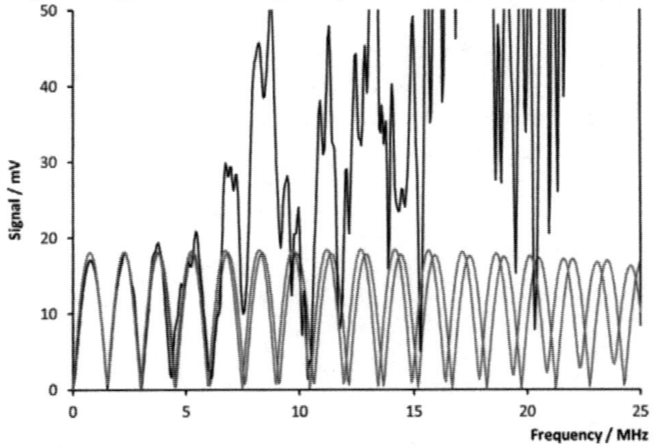

Figure 3 Frequency domain response for a slab of cavities of 10 μm radius for the effective medium model (red) and integrated ensemble average model (blue), and the discrete scatterer model at 2 v/v% (scaled by concentration).

3.3. Deviation from coherence

In order to investigate the conditions for the emergence of the coherent field from the incoherent response of a single realisation of scatterers, we quantify the degree of incoherence using the frequency domain response, such as that shown in Figure 3. As the measure of the degree of incoherence, we have adopted the sum of the squared residuals (RSS) between the discrete scatterer response and a reference response, taken to be either the effective medium (coherent) or ensemble average response. The sum is taken up to each frequency in turn. The RSS is normalised by cumulative sum of the square of the reference response. Further details are provided in [7]. A higher value of RSS indicates a less coherent response. The RSS value increases strongly with frequency (or bandwidth), and is also higher for larger cavity radii (at constant volume concentration), and for lower concentrations; these trends in the degree of coherence were previously observed in the time-domain responses of Figure 2.

The trends in the degree of coherence were established quantitatively by specifying a maximum value of RSS (relative to the effective medium response) which produced a coherent response; this value was estimated by assessing which time-domain responses were acceptably coherent. This maximum value of RSS=0.3 was then used to determine the maximum bandwidth (frequency) permitted to operate within this limit, for each condition of radius and concentration. The criterion is thus equivalent to specifying the maximum frequency at which a coherent response could be expected, for a single realisation of scatterers, for the particular cavity radius and concentration. Figure 4 shows how the maximum frequency varies with cavity radius and concentration for a single realisation at each condition. At larger cavity radius, a lower operating frequency is required to obtain coherence at the same concentration. Conversely, a lower frequency is required when working at lower concentrations of cavities at the same radius.

Power-law fits of the maximum frequency, f_{max} as a function of radius, a, and volume fraction, ϕ (each for a single realisation of the discrete scatterer model) were determined according to the relation

$$f_{\max} /\text{MHz} \approx D\left(a/\mu m\right)^{\alpha} \phi^{\beta} \tag{6}$$

where the fitted values are α=-0.71, β=0.32 with a constant factor D=36.7 (in appropriate mixed units). These fits are shown by coloured lines on Figure 4. It might be expected that the limit for coherent response would be related to the magnitude of the wavelength, compared with the mean distance between scatterers, as a measure of the typical length scale of heterogeneity in the system. This is equivalent to a coherence limit defined by the number density of scatterers per cubic wavelength. This was tested by an alternative power law fit (shown in black on Figure 4) where

$$f_{\max} /\text{MHz} \approx \left(G/\text{ms}^{-1}\right)\left(a/\mu\text{m}\right)^{-1} \phi^{1/3} \tag{7}$$

which resulted in a fit of G=335.1 ms^{-1}, corresponding to around 8 scatterers per (linear) wavelength. This functional dependence, however, does not appear to fit the data so well, as is clear on the plots, especially at small radius and high concentration.

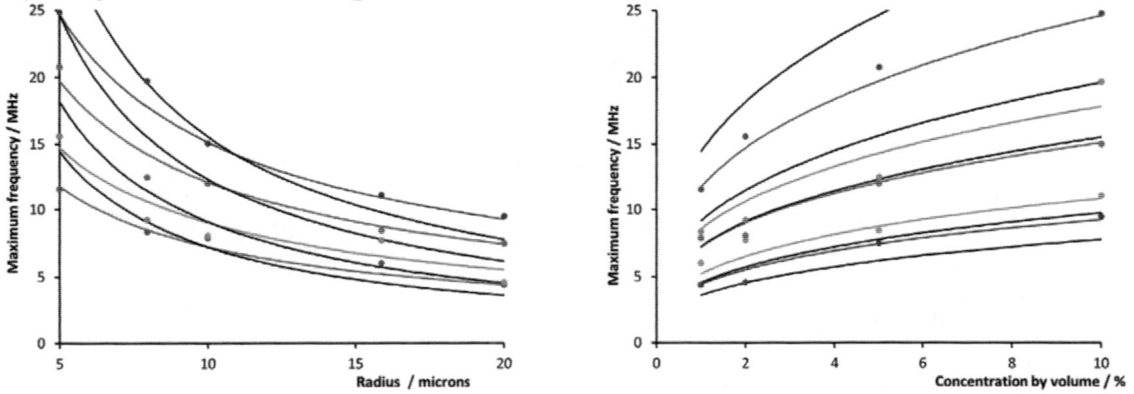

(a) (b)

Figure 4 Maximum bandwidth plotted against (a) radius and (b) concentration of cavities (a) concentrations of [1,2,5,10 %] are represented by blue, green, red, purple symbols respectively (b) radii of [5.0,7.9,10.0,15.9,20.0 μm] represented by blue, green, red, light blue, purple. The coloured fitted lines are given by equation (6), and the black fitted curves are for the fit on number density in equation (7).

3.4. Variability of incoherence

Thus far, the discrete scatterer results shown here have been for only a single realisation of scatterer locations. Since the configuration of scatterer locations is random, a different field response will be obtained for each new realisation, for a fixed cavity radius and concentration. The discrete scatterer model response averaged over many realisations will of course be the coherent field, since it that is an ensemble average response. However, each of the individual discrete scatterer model responses will be incoherent, and deviate from the coherent field by a different degree. By taking many realisations, we obtain the mean degree of incoherence, and the spread of that quantity. This has been determined using 15 simulations of the discrete scatterer model for a cavity radius of 10 μm for each concentration. The sum of squared residuals (RSS) was calculated between each discrete scatterer response (scaled by concentration) and the ensemble average response. Then, the mean and standard deviation of the RSS were calculated. The results are shown in Figure 5 as a function of frequency for the various concentrations; the curves are considerably smoothed compared with the results from any single realisation. Figure 5 shows the mean +/- one standard deviation shown by error bars, demonstrating the high degree of variation in the RSS value for different realisations of the same system. The RSS relative to the effective medium response (true coherent response) are similar, but include the error due to the effective wave speed which is not accounted for in the discrete scatterer and ensemble average models. Due to the spread of RSS values under the same conditions (radius, frequency, concentration), the definition of a coherent response condition would therefore require a confidence limit specification based on the standard deviation, to ensure that any single realisation would be likely to fall within an acceptable coherence level. At the approximate coherence condition obtained here, the standard deviation in RSS is around half the mean. Thus any definition of the coherence condition must consider the likely range of single realisation responses.

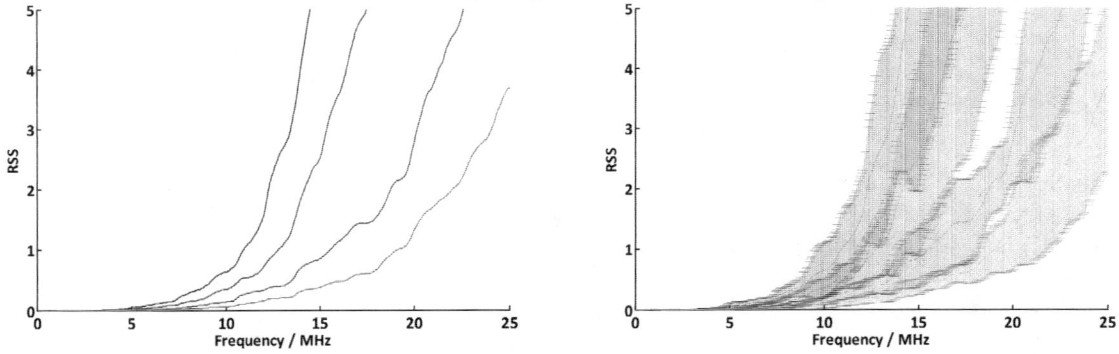

(a) (b)

Figure 5 The mean sum of squared residuals (RSS) as a function of frequency (bandwidth) for 10 μm radius cavities for concentrations of [1,2,5,10 %] shown in [blue, green, red, light blue] respectively; (a) mean of 15 simulations (b) mean+/- one standard deviation.

4. Conclusions

We have shown the results of simulations of the back scattered acoustic response from a slab region containing cavities in a solid matrix. The simulations have shown the emergence of the coherent response for even a single realisation of scatterer locations, under certain conditions. As expected, coherence results where there is small radius, higher concentration and lower frequency (longer wavelength). The degree of incoherence has been quantified and used to determine a maximum frequency at which coherence could be expected for a single realisation of scatterers for specific cavity radius and concentrations. The dependence on radius and concentration was compared with the criterion that the wavelength should be large compared with the average interparticle spacing; the fit showed similar trends but was not quantitatively accurate. We have also shown the significant variability in the degree of incoherence between multiple realisations of the same system (radius, concentration). Different realisations can produce quite different degrees of coherence. Although the model includes some simplifications, it has enabled some interesting features of the response of single realisations of scatterers to be explored.

References

[1] Foldy L L 1945 The multiple scattering of waves *Phys. Rev.* **67** 107-19
[2] Fikioris J G and Waterman P C 1964 Multiple scattering of waves II. "Hole corrections" in the scalar case *J. Math. Phys.* **5** 1413-20
[3] Lloyd P and Berry M V 1967 Wave propagation through an assembly of spheres IV Relations between different multiple scattering theories *Proc. Phys. Soc., London* **91** 678-88
[4] Linton C M and Martin P A 2006 Multiple scattering by multiple spheres: A new proof of the Lloyd- Berry formula for the effective wavenumber *SIAM J. Appl. Math.* **66** 1649-68
[5] Parnell W J and Abrahams I D 2010 Multiple point scattering to determine the effective wavenumber and effective material properties of an inhomogeneous slab *Wave Random Complex* **20** 678-701
[6] Pinfield V J and Challis R E 2011 Acoustic scattering by a spherical obstacle: Modification to

the analytical long-wavelength solution for the zero-order coefficient *J. Acoust. Soc. Am.* **129** 1851-6

[7] Pinfield V J and Challis R E 2012 Simulation of incoherent and coherent backscattered response of cavities in a solid matrix. *J. Acoust. Soc. Am.* **132** 3760-3769

[8] Ying C F and Truell R 1956 Scattering of a plane longitudinal wave by a spherical obstacle in an isotropically elastic solid *J. Appl. Phys.* **27** 1086-97

[9] Challis R E, Povey M J W, Mather M L and Holmes A K 2005 Ultrasound techniques for characterizing colloidal dispersions *Rep. Prog. Phys.* **68** 1541-637

11th Anglo-French Physical Acoustics Conference (AFPAC 2012) IOP Publishing
Journal of Physics: Conference Series **457** (2013) 012010 doi:10.1088/1742-6596/457/1/012010

Passive impact localisation for the structural health monitoring of new airframe materials

Jacob Ward[1,2], Anthony Croxford[1] and Christophe Paget[2]

[1] Ultrasonics and non-destructive testing group, Department of Mechanical Engineering, University of Bristol, BS8 1TR, UK

[2] Airbus in the UK, New Filton House, BS99 7AR, UK

E-mail: jake.ward@bristol.ac.uk

Abstract. This experimental work considers the use of permanently attached sensors for the detection and location of impacts to a carbon fibre reinforced plastic panel with stringers. Deterministic knowledge of the propagation of Lamb waves in the structure is not used. Instead a statistical measure of the signal is used to determine the arrival time of elastic waves propagating in the structure as a result of the impact. A comparison is made between a conventional method and the statistical method. The conventional method, which has been routinely used in industry for acoustic emission imaging, uses the timing of a peak in the recorded signal. The statistical method uses the Rayleigh maximum likelihood estimator. The statistical method is shown to provide both more precise and robust estimates of the elastic wave arrival time. An array of just four sensors is used to locate the impacts. The accuracy of the localisations is used to visualise the effectiveness of the two methods for the low sensor density used. Low sensor density is necessary for minimising system weight and cost. The equivalent net sensor density used in this experiment was five sensors per meter squared. Carbon fibre reinforced plastic is today used for both exterior surfaces and primary structure of airframes entering service. The industrial relevance of this work is to mitigate the diminishing role of visual inspection for evaluating the health of aerospace structures, where impact damage may not be visible.

1. Introduction

This paper presents an experimental study into the localisation of impacts on a composite panel using a small number of sensors. The industrial relevance of the technique is due to new materials being selected in airframe design. Historically airframe structures have largely comprised aluminium alloy components. The aerospace industry and operators have consequently been heavily reliant on visual inspection as a means of detecting, for example, fatigue cracking or impact damage because it can normally be seen in these materials. Much recent design effort at Airbus has been spent on increasing the percentage of carbon fibre reinforced plastic (CFRP) in both civil and military products. The new military airframe A400M, as an example, widely utilises CFRP material on exterior surfaces. CFRP material fails in a different way to aluminium alloy, particularly under impact. Impact damage to CFRP material cannot always be seen and because of this the effectiveness of visual inspection, as a means of non-destructive evaluation (NDE), is set to reduce in the future.

For the ageing fleets of metallic airframes currently in service, the visibility of impact damage is an important factor in its detection. This is due to the predictable way aerospace alloys yield and fail. Visual inspection forms part of routine maintenance of these aircraft and when

Content from this work may be used under the terms of the Creative Commons Attribution 3.0 licence. Any further distribution of this work must maintain attribution to the author(s) and the title of the work, journal citation and DOI.
Published under licence by IOP Publishing Ltd

significant damage is found it is typically entered into the airframe logbook and an in-depth evaluation would be commissioned. An in-depth evaluation of impact damage requires suitable NDT techniques to be used. Such inspections require skilled engineers and specialist equipment to be brought to the structure and typically incurs substantial cost to the operator.

One important risk in migration to advanced CFRP composite material is that impact damage may be missed due to its poor visibility and consequently grow through fatigue. Barden and Almond et al.[1] compared optical and thermal methods as candidate screening tools for rapid evaluation of large structural areas of CFRP composite material, detailing the risk of barely visible impact damage. An alternative method to mitigate such risk would be to automatically monitor and locate structural impacts by measurements made from permanently attached sensors. This technique would use measurement of structure-borne elastic waves. Impacts to solid structures can be located passively by measuring time-of-arrival of resultant propagating waves at remote sensor locations. The precision of the time-of-arrival measurements determining the accuracy of impact localisation.

Firstly this paper discusses a conventional approach to measuring time-of-arrival which has routinely been used for acoustic emission structural health monitoring (SHM). Acoustic emission is the passive measurement of ultrasonic frequencies arising from brittle fracture events measured from sensors placed away from the fracture site. Tobias presented a method to localise sources in two dimensions using time difference of arrival from three sensors[2]. Triangulation techniques often applied to acoustic emission data fundamentally rely on measuring time-of-arrival for the purpose of locating sources. This can be particularly complex in anisotropic media[3]. Two common methods for measuring time-of-arrival are borrowed from acoustic emission techniques and are discussed for their relevance to impact localisation. This paper highlights difficulties with the common methods and show them to be unfit for purpose, often due to erroneous or unreliable localisation. This paper proposes an alternative signal processing method for time-of-arrival measurement taken from a recent publication on active SHM[4]. The alternative is found to improve the precision and reliability of time-of-arrival measurements leading to a convincing improvement in the accuracy of impact localisation when compared to conventional methods applied to the same data. The discussion section includes practical aspects of implementation and an interpretation of the results.

2. Method, equipment and materials

2.1. Background to a conventional approach

An extensive literature on SHM has developed over the past 25 years. The sequence of detection, localisation and characterisation of damage were all achieved by acoustic emission in the experimental work of Scruby and Baldwin et al.(1985)[5]. This was before the principles of SHM were defined, however the same sequence is generally still in use today[6]. The passive technique of acoustic emission continues to receive attention from manufacturers in the aerospace sector[3][7] and the overarching principles of SHM have been elaborated in the review work of Worden and Farrar et al.[6]. The method in this paper considers the problem of source localisation in isolation from other problems of detection and identification, which are not covered. The method applied is for localising structural impacts, however, the method bares relevance to acoustic emission source localisation and takes standard practice from this discipline as a benchmark for the alternative proposed.

The passive method of acoustic emission monitoring is one SHM method that has successfully permitted the localisation of sources from remote measurements at ultrasonic frequency. The sources of interest in acoustic emission measurement are typically brittle fracture events in metallics and other materials resulting from loading. The modelling work of Wilcox and Lee et al.[8] showed that source location and threshold level were critical factors affecting the probability of detection at either any or all sensors in an array. Acoustic emission monitoring equipment has been developed by a consortium that included Airbus. The equipment uses a network of charge-amplified piezoceramic sensors. This equipment has been made airworthy to technology readiness level six by an extensive programme of testing. The equipment was developed for over a decade by a consortium comprising Lloyds Register, Ultra-Electronics and Airbus[3][7][9]. One key capability offered by the equipment is to communicate information for locating ultrasonic sources using a minimal transmission of data. Algorithms developed made progress on source localisation in anisotropic material, accommodating the anisotropic wave-propagation resulting from carbon fibre layup. Numerical studies such as Torres-Arredondo et al.[10] provide insight into the complex behaviour of wave propagation in these materials and show several ways to visualise it. Once anisotropy has been characterised, either empirically or numerically, there are methods available to compute source location from arrival times affected by the anisotropy of propagation[3][11].

Localisation algorithms, such as Paget and Atherton et al.[3], use time-of-arrival data to triangulate sources, however, one set back has often been the method in which time-of-arrival has been measured. Historically time-of-arrival has been obtained by threshold-crossing times or peak arrival times. The measured signal is normally rectified by the acoustic emission hardware, as shown in Figure 1. The purpose of applying either of these standard techniques being to measure the relative time-of-arrivals between different sensors in the array. The two standard measurement methods can be carried out with simple cheap light-weight electronics, making them attractive methods for applications in aerospace. However, both of these techniques are fundamentally flawed and are flawed for different reasons.

An example of an impact signal is given in Figure 1. The preprocessing step of rectifying the pre-amplified signal is achieved using diodes. The timing of a 0.4 Volt threshold-crossing and the detection of an amplitude peak have been marked on the signal using coloured cross-hairs.

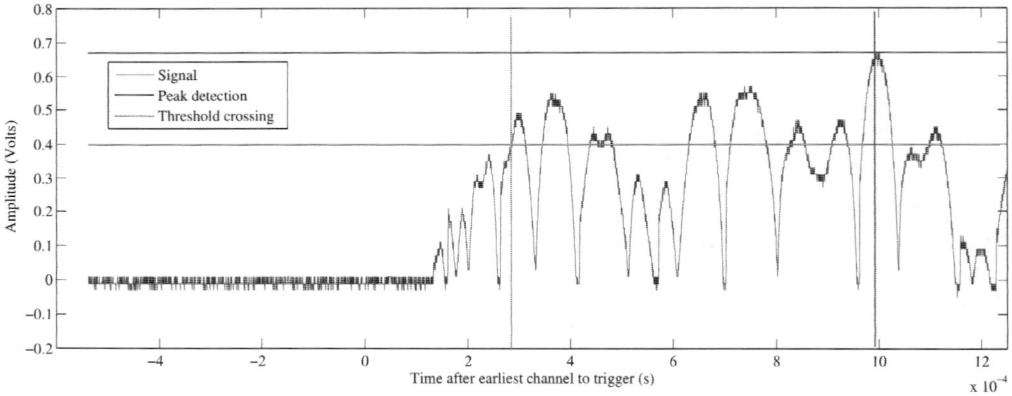

Figure 1. A threshold crossing time, or peak time, are routinely used for measuring the time-of-arrival of propagating elastic waves for industrial acoustic emission applications

2.2. Difficulties with the conventional approach

Threshold crossing is flawed because Lamb wave dispersion and divergence causes delays in the timing and, furthermore, the threshold level needs to be set relative to ambient background vibration levels, which may change according to environmental conditions of operation. In a study of acoustic emission measurements applied to landing gear structures, el-Bakry showed that a 6dB amplitude change could lead to 25μs difference in time-of-arrival measurement using the threshold crossing technique on 300kHz-band sensors applied to a 300M steel plate[7]. The peak-detection method was introduced to overcome some of the basic problems associated with threshold-crossing. The peak-detection method requires an arbitrary time-out parameter to be set which defines when the peak-hold circuit is reset, at which point the timing of a peak can be transmitted and a new peak detected. Like threshold-crossing, the peak-detection method can be implemented using simple light weight analogue circuits. However, the peak-detection method is flawed for a different reason. The reason the peak-detection method is flawed is because the reverberant field from a source can lead to higher recorded amplitudes than the original direct path of elastic waves. This is illustrated by the example given in Figure 1 where the blue cross-hairs marking the peak occurs significantly late in the time history. This can arise from either multiple Lamb wave modes propagating at different velocities or the transient interference pattern of waves reflected from large coherent reflectors such as stringers or edges. The uncertainty around the timing of the peak has been found to be a particular problem when applying the peak-detection method to metallic panels which have relatively low levels of visco-elastic absorption compared to CFRP composite. Peak-detection is consequently the standard method used for the acoustic emission testing of CFRP structure and therefore taken as the benchmark for the study here into impact localisation.

2.3. The proposed RMLE method as an alternative

A new approach is taken in this experimental work for estimating the time-of-arrival of propagating elastic waves resulting from structural impact. This approach adopts a technique that has recently seen attention for active guided wave SHM and uses the Rayleigh maximum likelihood estimate (RMLE) of parts of the signal[4].

The useful property of the RMLE method in application to source localisation is that an apex obtained at the time-of-arrival of the elastic waves. The RMLE method differs from the threshold-crossing method because there is no need to set a threshold parameter relative to the ambient structural vibrations or fretting noises featuring in the measurement. The new method uses a statistical treatment of the sampled signal rather than any direct interpolation from the rectified signal itself. The RMLE method has been termed the arrival filter in Flynn and Todd et al. where a full derivation can be found[4].

The RMLE method uses a variable sample index, η, to divide the signal into two parts. The samples either side of the division point have different amplitude distributions. The maximum likelihood estimates for the two regions of the signal, Equations 1 and 2, are shown as a function of the division point η, and the full arrival filter output $\omega[\eta]$ can be computed using Equation 3.

$$\sigma_1(\eta) = \sqrt{\frac{1}{2(\eta)} \sum_{n=1}^{\eta} \nu^2[n]} \tag{1}$$

$$\sigma_2(\eta) = \sqrt{\frac{1}{2(N-\eta)} \sum_{n=\eta+1}^{N} \nu^2[n]} \tag{2}$$

$$\omega[\eta] = -\eta \ln \sigma_1^2(\eta) - (N-\eta) \ln \sigma_2^2(\eta) + 2 \sum_{n=1}^{N} (\ln(\nu[n])) - N \tag{3}$$

where N is the number of points in the series and ν is the scalar magnitude of the sampled time series, which for computational reasons cannot be exactly zero at any point due to the logarithm used for calculation. It can be seen that the arrival filter is non-linear and acausal, however, this is not strictly a problem as many SHM systems operate in post-processing rather than in real-time [12].

2.4. The experiment

An experiment was designed to compare the arrival filter (RMLE method) with the standard method of peak detection. A photograph of the specimen used for the experiment is given in Figure 2. The CFRP panel containing two stringers was impacted at a number of locations. The impacts were provided by dropping a 10mm diameter steel ball bearing down a 30mm diameter tube from a 200mm height. A diagram is shown in Figure 3. The spacing between impact locations was 100mm. The impacts were to the exterior surface of the lower wing cover specimen, to the reverse of the stringer-side shown in Figure 2. Ten impacts were carried out at each of the 42 locations. Knowledge of the true impact location was therefore correct to $\pm$ 15mm of the target impact site.

Figure 2. The specimen used for the experiment was a CFRP composite panel containing two stringers

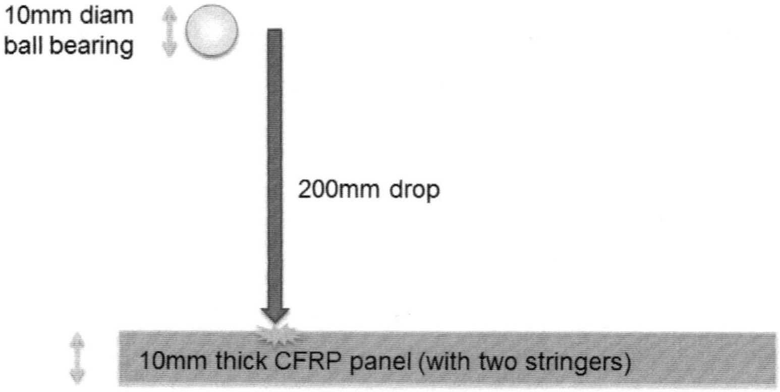

Figure 3. The experiment used a ball-bearing drop to provide an impact source

Four sensors were used for the localisation of impacts over a 7×6 grid as shown in Figure 4. The array transducers and pre-amplifiers were centred on 7kHz frequency, significantly lower than typical acoustic emission measurement frequencies. The area covered by the array was 0.6×0.6 metres. If these array elements were to tessellate over a larger structure in a regular square grid with the arrangement shown, this would correspond to a net sensor density of 5.5

sensors per metre squared. The time-of-arrivals were taken from all four channels during the experiment using both standard and new method.

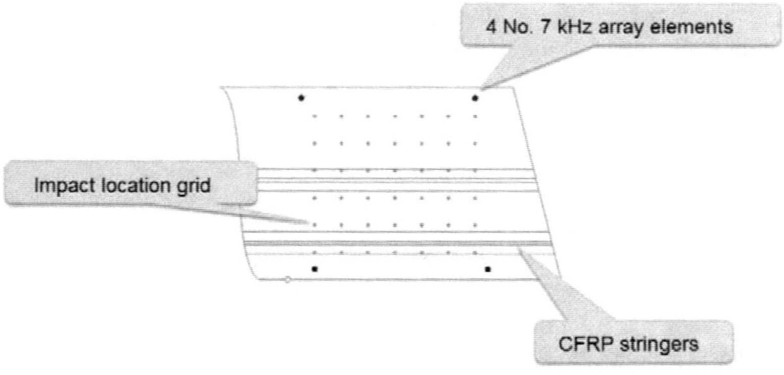

Figure 4. A grid of 42 impact locations were used

A significant problem with passive methods is that the impact event-time is not known. Only the relative differences in arrival times can be deduced from synchronous recordings at sensor positions. This is not a problem encountered with active imaging, because the time of transmission will generally be known using this method. The important difference between active and passive array equipment affects the electronic architecture that can be used, ultimately affecting cabling weight. The data from an active array can be built up sequentially by transmitting and receiving on different pairs each time. Active SHM networks therefore use multiplexing to enable a limited number of transmit/receive channels to cover a large number of elements. The transmitted signal can be repeated as many times as necessary to reduce random noise affecting the measurement, allowing low power excitation to be used. Michaels and Lee et al.[13] cite signal averaging as a common method to improve signal to noise ratio on active networks. They note this unfortunately leads to prohibitively long acquisition times when multiplexing over arrays larger than 10 elements. The alternative to averaging presented in their work uses pulse compression from recorded responses to chirped excitations, consequently speeding up the acquisition process. Whereas the active SHM network is able to use multiplexed channels, the passive technique intrinsically needs synchronous multi-channel recording. This emphasises the need for low sensor density particularly on passive techniques to avoid prohibitively high cabling weights.

The triangulation algorithm used in this paper is designed for localising sources in plates where wave-fronts propagate elliptically as a result of anisotropy. This kind of anisotropy might be expected in CFRP plate comprising unidirectional fibres. The intersection of two ellipses can be solved analytically as the real roots of a quartic equation using knowledge of two different group velocities in orthogonal directions. A full derivation can be found in Paget and Atherton et al.[3]. For this experiment these were measured empirically as $c_{gr}^{0^\circ} = 1560ms^{-1}$ and $c_{gr}^{90^\circ} = 1390ms^{-1}$ respectively using threshold-crossing. $c_{gr}^{0^\circ}$ is the effective group velocity in the predominant fibre direction, which in this case was parallel to the stringers and $c_{gr}^{90^\circ}$ is the group velocity perpendicular to stringers. This algorithm differs from that given in Scholey and Wilcox et al.[11], where the full angular dependence of group velocity must be obtained to compute likely source location. Whereas the method in Paget and Atherton et al. requires arrays to be of three or four sensors specifically, the method of Scholey and Wilcox et al. can be applied to an arbitrary number of sensors.

The aim of this work is not to evaluate triangulation method, but to evaluate the performance and reliability of time-of-arrival estimates. The triangulation is however helpful in presenting the large amounts of data collected during the experiment and gives a practical example of achievable accuracy.

3. Results

An example of the time-of-arrival measurements achievable with the two methods is given in Figures 5 and 6. Figure 5 shows the rectified signal from three sensors with the timings of the peaks from three channels marked by vertical blue lines. Figure 6 shows the RMLE method applied to the same signals shown in Figure 5. The time-of-arrivals are taken from the maxima in the RMLE output. The times at which these occur are indicated here by vertical red lines. The values interpolated for the two methods are given in Table 1.

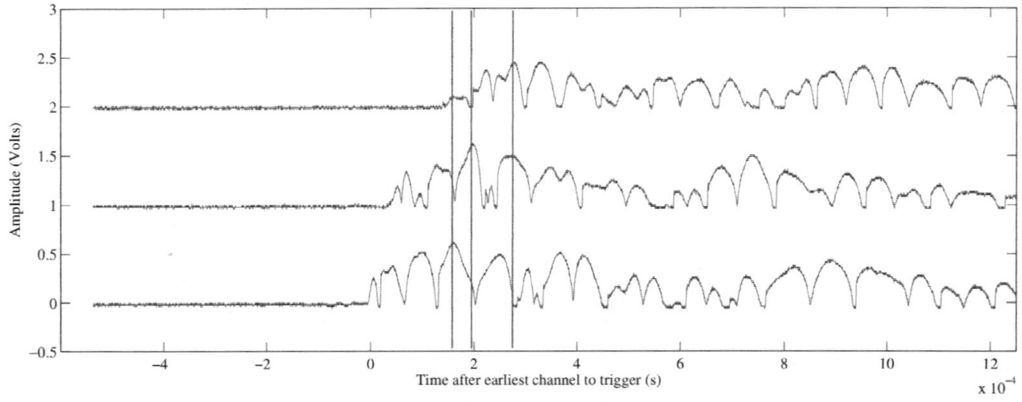

Figure 5. Measured times-of-arrival using the peak-detection method

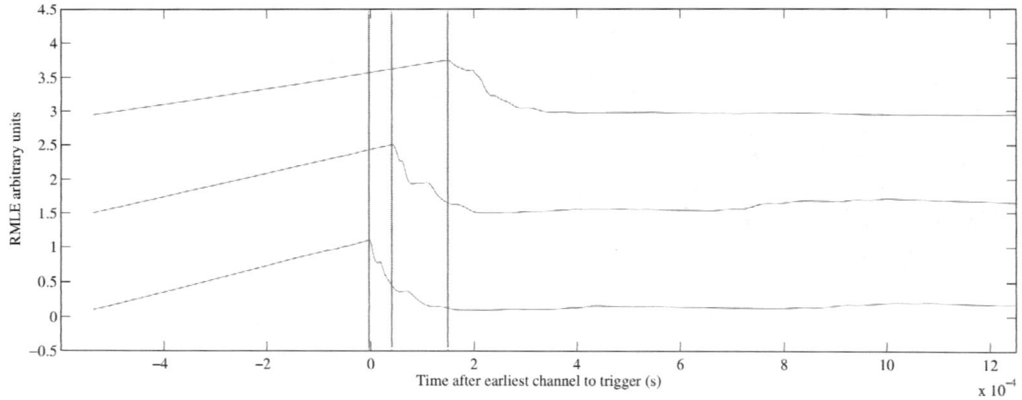

Figure 6. Measured times-of-arrival using the Rayleigh maximum likelihood method

Table 1. Time-of-arrival estimates produced by the peak and RMLE methods from examples given in Figures 5 and 6. All times in μs.

Channel	*Peak*	*RMLE*	ΔT_{peak}	ΔT_{RMLE}
1	158.40	−3.20	—	—
2	195.20	40.40	36.80	43.60
3	274.00	149.20	78.80	108.80

It can be seen from Figures 5 and 6 that the RMLE method measures the arrivals at significantly earlier times than the peak-detection method. There is also a subtle difference in the time difference of arrival, seen by the horizontal spacing between the lines. The timings output from the two methods are summarised in Table 1 for a single impact event. It can be seen from the RMLE arrival time shown on Channel 1 that oscilloscope trigger is pre-empted by -3.2μs. This corresponds to the arrival time at first-hit sensor, which would ideally be time zero, and demonstrates the sensitivity of the technique to the "true" first arrival time.

The times-of-arrival given by the two methods (Figures 5 and 6) were used to localise the 10 impacts at each of the 42 impact locations. The spread of localisations from the two methods are given in Figure 7. Figure 7 shows there is a greater spread in the localisations produced by the time-of-arrivals obtained using the peak-detection method than the RMLE method. The poorest performing localisations obtained using the peak-detection show extremely large error, to the extent that it cannot obviously be seen which localisations relate to the which impact sites.

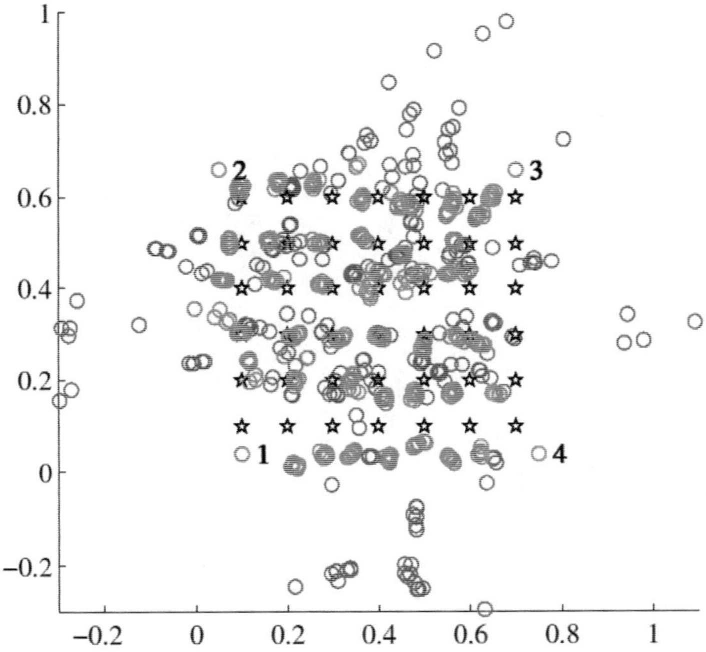

Figure 7. All impact location estimates using the peak-detection (blue) and RMLE (red) methods. Transducer positions shown in green and impact locations in black

These localisation results for both methods have been superimposed onto the surface that was impacted (Figure 8) and the reverse side of the panel (Figure 9) using image registration. Figure 9 reveals the stringer locations relative to the impact localisations. The red circles in Figure 9 show there is a greater extent of error near the bottom and right hand sides than in other areas of the panel using the RMLE method. This could perhaps be related to the layout of sensors in relation to stringer positions, which can be seen more clearly in Figure 9.

The magnitude of the localisation errors are plotted in Figure 10 for each of the ten repetitions at all impact locations. This gives an idea of the spread of results when the impact is repeated using the two techniques. The green line at the bottom of Figure 10 indicates the $\pm$ 15mm tolerance on true impact location as mentioned in Section 2.4. This arises from insufficient control of the actual impact site. Firstly the figure shows that the errors from the peak-detection method are far greater in magnitude than the RMLE method for almost all impact sites. The second thing to notice from Figure 10 is that the variance between impact localisation errors is also greater for the peak-detection method than the RMLE method at each impact site. This highlights one key issue of the peak-detection method being that of extremely high sensitivity towards precise impact location. The RMLE method, in contrast to peak-detection, shows a

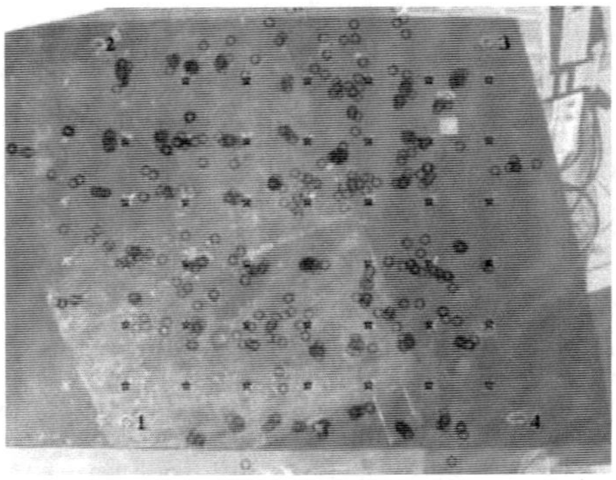

Figure 8. Impact localisation superimposed onto the external surface of the CFRP wing panel specimen

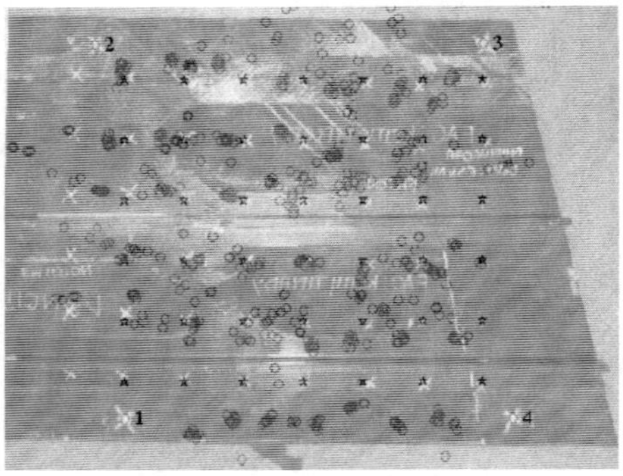

Figure 9. Impact localisation superimposed onto the internal surface (photograph flipped left-right)

more robust and stable localisation from 10 repetitions. The higher error and variance from the peak-detection results making the RMLE method more favourable.

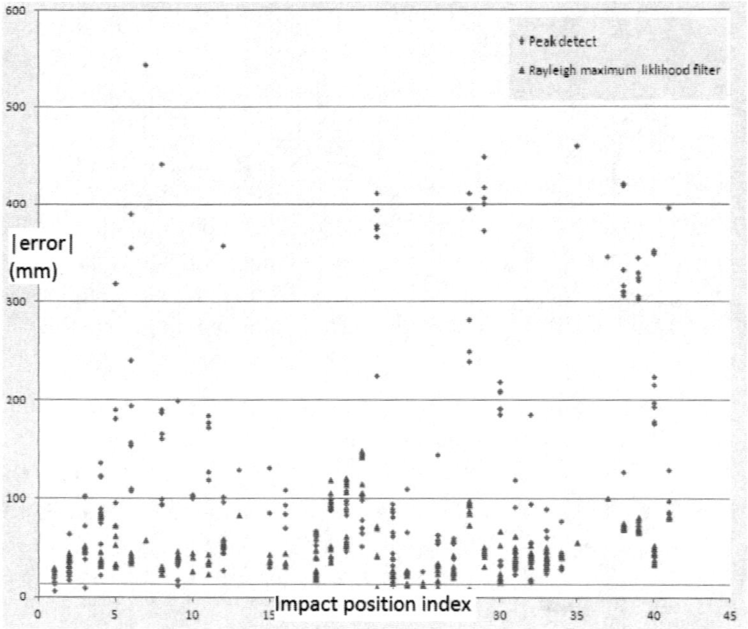

Figure 10. Errors shown for the ten repetitions of measurements at the 42 impact locations

4. Discussion

4.1. Practical aspects of implementation

The RMLE technique is sensitive to changes in the signal distribution over time, rather than the sampled signal itself. Unlike threshold-crossing it does not require a threshold to be set above noise floor. Unlike peak-detection it does not require a time-out to be set, after which higher peaks cannot change the time-of-arrival. By removing input parameters such as these,

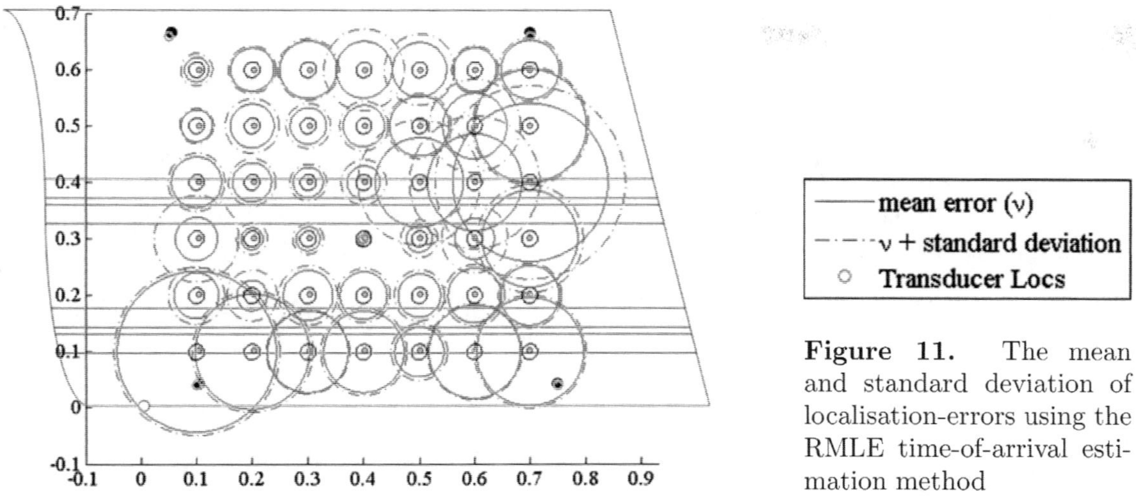

Figure 11. The mean and standard deviation of localisation-errors using the RMLE time-of-arrival estimation method

there are fewer obstacles for the technology to be implemented in industry. It may also be that the RMLE method is more robust against benign structural vibrations than the conventional methods, however, this would require further study.

One challenge affecting the deployment of the RMLE technique is that it requires the whole time history to be post-processed digitally, rather than using the analogue circuits available from acoustic emission methods. The RMLE method also requires a pre-trigger of signal prior to impact. Combining this with the need to capture pre-trigger on multiple channels, this affects data gathering requirements and therefore cost and weight.

4.2. Interpretation of results

If impact localisation is required as part of SHM, the difficulties of the conventional methods need to be overcome. The difficulties with the conventional approach were outlined in Section 2.2. These arise because impacts generate a superposition of many propagating and dispersive Lamb wave modes, making time-of-arrival difficult to estimate. It has been shown that this leads to poor localisation accuracy when using peak-detection and that the RMLE method shows significant improvement. Figure 10 shows greater error and lower consistency using the peak-detection method. The size of the error for the peak-detection method being comparable to the size of the plate specimen on many localisations. For this reason the peak-detection results are omitted from Figure 11 and, by contrast, the RMLE method attributes source location to an accuracy within the 100mm grid spacing of the impact sites. A large standard deviation on impact localisations can be seen in some areas of Figure 11. This can be attributed to a tolerance on actual impact location as indicated by the green line in Figure 10 relative to the impact location grid (Figure 4). The control over the actual impact site was not fine enough to consistently hit either above the stringer-foot area or open plate between the ten impact repetitions. This is thought to explain the results on the centre right hand side of Figure 11. Of the ten impact trials at this location it can be expected that the panel was impacted above the stringer-foot only on some occasions, leading to variance in the result. This is unlike the lower left hand side where the error was equally large but variance small.

5. Conclusion

The RMLE technique can locate impact when it occurs by use of permanently attached equipment. The new technique has been shown to give more repeatable time-of-arrival measurement using ball-bearing impact than conventional methods historically implemented on acoustic emission testing. More precise time-of-arrival measurement led to more accurate localisation using a small number of sensors. Impacts may be detected by threshold crossing, however, neither threshold crossing nor peak-detection should be used for impact localisation in large complex CFRP structures.

The reliability of electronics and sensor network are central to the business case for SHM technology. It is an open question whether digital equipment for post-processing can be made as reliable as previously used analogue circuits for acoustic emission testing. Any high costs of maturing this technique should be considered carefully against the changing demands of inspection, in particular the diminishing role of visual inspection in the long term.

This paper used source localisation accuracy as a performance metric for the purpose of evaluating arrival time measurement. A quantitative study into the precision of arrival times would be useful progress to this work, however, care should be taken when choosing the "true" arrival time for benchmarking, because the correct value is not necessarily clear. One practical question of importance is how many samples of the signal are necessary to enable the RMLE technique to work.

Acknowledgments

The Engineering and Physical Sciences Research Council
The United Kingdom Research Centre in Non-destructive Evaluation (RCNDE)
Airbus in the UK (Filton)

References

[1] Barden T, Almond D, Pickering S, Morbidini M and Cawley P 2007 *Detection of impact damage in CFRP composites by thermosonics* (NDT and E, Vol. **22(2)**) pp 71–82
[2] Tobias A 1976 *Acoustic-emission source location in two dimensions by an array of three sensors* (Non-Destructive Testing) pp 9–12
[3] Paget C, Atherton K and O'Brien E 2003 *Triangulation algorithm for damage location in aeronautical composite structures* (Proceedings of the fourth international workshop on structural health monitoring, DEStech Publications Inc.)
[4] Flynn E B, Todd M D, Wilcox P D, Drinkwater B W and Croxford A J 2011 *Maximum-likelihood estimation of damage location in guided-wave structural health monitoring* (*Proc. Roy. Soc.* A-mathematical physical and engineering sc, Vol. **467** (2133)) pp 2575–96
[5] Scruby C B, Baldwin G R and Stacey K A 1985 *Characterization of fatigue crack extension by quantitative acoustic emission* (International Journal of Fracture Vol. **28 (4)**) pp 201–22
[6] Worden K, Farrar C R, Manson G and Park G 2007 *The fundamental axioms of structural health monitoring* (*Proc. Roy. Soc.* A-mathematical physical and engineering sc, Vol. **463**) pp 1639–64
[7] el-Bakry M 2011 *Acoustic Emission Monitoring of Safe-Life Landing Gear Structures* (unpublished engineering doctoral thesis, Queens Building Library, University of Bristol, RCNDE) chap **3**: Acoustic Emission Detection and Location, pp 64–78
[8] Wilcox P D, Lee C K, Scholey J J, Friswell M I, Wisnom M R and Drinkwater B W 2006 *Progress towards a forward model of the complete acoustic emission process* (Advanced Materials Research Vol. **13-14**) pp 69–75
[9] Atherton K J, Paget C A and O'Brien E 2004 *Structural Health Monitoring Of Metal Aircraft Structures With Modified Acoustic Emission* (SEM X International Congress and Exposition on Experimental & Applied Mechanics)
[10] Torres-Arredondo M A, Jung H and Fritzen C 2012 *Towards the development of predictive models for the system design and modal analysis of acoustic emission based technologies* (Key Engineering Materials Vol. **518**) pp 396–406
[11] Scholey J, Wilcox P D, Wisnom M R, Friswell M I, Pavier M and Aliha M R 2009 *A generic technique for acoustic emission source location* (*Journal of Acoustic Emission* Acoustic Emission Group, Vol. **27**) pp 291–8
[12] Croxford A, Wilcox P, Drinkwater B and Konstantinidis G 2007 *Strategies for guided-wave structural health monitoring* (*Proc. Roy. Soc.* A-mathematical physical and engineering sc, Vol. **463**) pp 2961–81
[13] Michaels J E, Lee S J, Croxford A J and Wilcox P D 2012 *Chirp excitation of ultrasonic guided waves* (Ultrasonics, in press **DOI:** 10.1016/j.ultras.2012.06.010)

Numerical modelling for the optimization of multi-element, capacitive, ultrasonic, air-coupled transducer

D. Zhang, M. Rénier, M. Castaings

Univ. Bordeaux, I2M, UMR 5295, F-33400 Talence, France.
CNRS, I2M, UMR 5295, F-33400 Talence, France.
Arts et Métiers ParisTech, I2M, UMR 5295, F-33400 Talence, France.

Email: d.zhang@i2m.u-bordeaux1.fr

Abstract. This paper presents an analytical method to predict the acoustic pressure field produced in the air by concentric, annular sources. The sources correspond to a multi-element, ultrasonic, air-coupled array, which should be used further for non-destructive testing (NDT) purposes. After validation of the model based on the comparison of the predicted acoustic pressure with experimental results, a numerical study of the consequence of the annular sources arrangement on the acoustic field is realized. This helps in optimizing the design of the array, so that the intended ultrasonic field is produced. For high spatial resolution to be obtained in further NDT process this field should be focused and, ideally, the frequency used for inspecting the tested piece should be tuned so that wavelengths in the inspected material take suitable values in regards of piece thickness and defect sizes. Numerical simulations show that the multi-element array allows dynamic focussing to be made by judiciously adjusting delay laws of the excitation signals applied to each individual element.

1. Introduction

The application of ultrasounds in both biomedical imaging and industrial non-destructive testing (NDT) has been greatly advanced recently. Air-coupled techniques for the generation-detection of ultrasonic waves in tested materials are very advantageous because they avoid the use of coupling liquids as gel (contact technique) or water (jets or immersion tanks) for instance, which render much more complicated online testing procedures. Today's available air-coupled techniques for sending/receiving ultrasounds to/from tested samples are based on the use of either specific ultrasonic transducer [1][2], laser sources and/or probes [3] or electromagnetic acoustic transducers (EMATs) [4]. Laser-based techniques are quite expensive and rarely appropriate for on-line inspection because they require high level of accuracy in measurement set-up; EMAT-based techniques are suitable for conductive or magnetostrictive components only, thus limiting the field of applications. Consequently, as an alternative economical solution, air-coupled ultrasonic transducers could be used for the inspection of any metallic or polymer-based material pieces. However, one difficulty arising when using air-coupled ultrasonic transducers is the huge energy transmission losses at each solid-air interface, such as those between the transducers and the air, or those between tested materials and the

Content from this work may be used under the terms of the Creative Commons Attribution 3.0 licence. Any further distribution of this work must maintain attribution to the author(s) and the title of the work, journal citation and DOI.
Published under licence by IOP Publishing Ltd

air, e.g. up to -54 dB loss at one air-aluminium interface [5]. Moreover, the attenuation of ultrasounds in air increases with the frequency according to the following law $1.64f^2 \times 10^{-10} dB/m$ [6], where f is the frequency in Hertz. For example, the amplitude loss for a 1 MHz frequency wave will be equal to about 85 dB after a distance equal to 50 cm. Consequently, for usual NDT applications, frequency must be kept down to reasonably low values, *i.e.* below 0.5 MHz, and transducers efficiency must be as large as possible. Furthermore, for NDT purposes, transducers should be of large frequency bandwidth to make frequency tuning possible, so that optimum frequency is set according to the thickness of the inspected piece and/or to the size of sought defects.

Capacitive ultrasonic transducers have been developed over the last two decades [7][8][9][10], and successfully used for the contact-less inspection of composite structures, for instance [11]. Their main advantages include a very large frequency bandwidth [12] and a sufficiently high sensitivity [13] for general non-contact NDT. Several manufacturing techniques have been reported, including series of coarse polishing [14] machining V-shaped grooves [15][16] or circular holes through laser ablation [8], photolithography [12], or sand blasting [9] so that obtained roughness or cavities at the surface of the metallic back-plate determine the frequency properties of the capacitive ultrasonic transducer [17]. Also, these elements require a few-micron thick membrane deposited on the back-plate, with metallised external surface. This must be stretched for the transducer to operate, thus producing locally plane wave front and introducing some difficulties in allowing the focusing of the ultrasonic beam, causing low spatial resolution, as a serious drawback for some NDT applications. In order to circumvent this problem, a method has been proposed using a micro-machined Fresnel zone-plate placed in the acoustic field [18]. A flexible back-plate as been developed as an alternative solution, allowing the realization of concave capacitive air-coupled transducers [19]. With micromachining fabrication technology on the silicon back-plate a micro-electromechanical system for cMUT (capacitive micromachining ultrasonic transducer) has been developed [20] recently. However, these solutions offer fixed focusing features and as many transducers as intended focusing conditions will be required for satisfying several NDT inspections.

The objective of this paper is to present a model and its use to optimize the design of ultrasonic air-coupled transducers. These should allow the generation of an ultrasonic beam with dynamic focusing properties, with a large frequency bandwidth and with high efficiency. The proposed solution is a multi-element, capacitive, array, so that the capacitive part may bring the large frequency range, and the multi-element feature will allow for dynamic focusing capabilities under appropriate phase delay control. This is made up of one central circular element and several concentric, annular elements. Two cases are studied, with different spatial configurations. In the first case, the difference between the outer and inner diameters is kept constant for every annular element. Thus, the active surface of the rings increases as rings get away from the axis of the assembly. In the second case, the elements have dimensions such that their respective active surface remains constant. In section 2, the analytical model, which is based on the Kirchhoff integral, is presented. It allows numerical predictions of ultrasonic fields radiated by either single or multi-element transducers to be made. Section 3 presents a laser-based, experimental set-up and its use for measuring the ultrasonic pressure field produced by a circular mono-element for validating the analytical model. Then the model is used for predicting the ultrasonic field produced by one single annular element, and experiments are preceded for validation purpose. Agreements and discrepancies are discussed. Finally, in section 4 the multi-element arrays are numerically simulated. The focusing properties of two configurations of element arrangement are compared, demonstrating that dynamic focusing is possible if adequate delay laws are used for exciting each element of the array.

2. Modelling the acoustic field pressure generated by air-coupled transducers

We consider the wave propagation of ultrasound in air radiated by an air-coupled transducer. The model lies on the assumption that the vibration of the transducer surface is spatially uniform. The expression uses Kirchhoff integral to calculate the pressure in the frequency domain $p(r,z;\omega)$. In order to save computer resources and computational time, the model is built as a 2D axial-symmetric model, under polar coordinate system (r, z) (**Figure 1**).

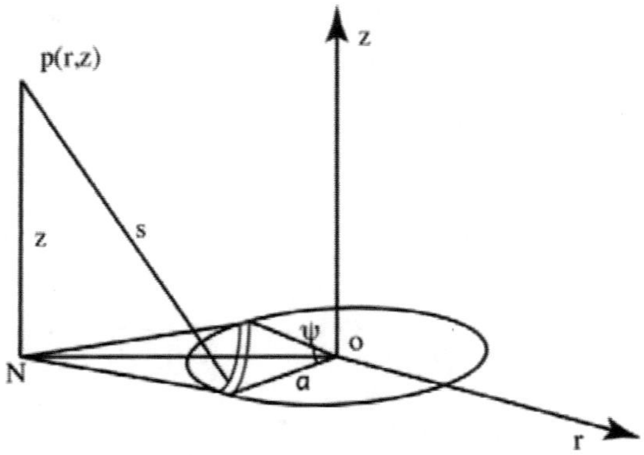

Figure 1. Definition of polar coordinate axes for the transducer acoustic field model.

Starting from a harmonic and uniform motion of the transducer surface, the normal velocity of the transducer surface expresses as $V(t)=v_0 e^{j\omega t}$, where v_0 is the uniformed amplitude of the normal velocity and ω is the angular frequency of the excitation signal applied to the transducer. The steady-state pressure at a fixed point in the medium $p(r,z;\omega)$ could be derived using the following equation:

$$p(r,z;\omega)=-j\omega\rho\tilde{v}H(r,z;\omega) \tag{1}$$

where $\tilde{v}$ is the Fourier transform of $V(t)$, ρ is the density of air; $H(r,z;\omega)$ is the Fourier transform of the impulse response $h(r,z;t)$, which could be derived as [21],

$$H(r,z;\omega)=\frac{ac_0}{j\omega\pi}\int_0^{\pi}\frac{r\cos\psi-a}{r^2+a^2-2ar\cos\psi}\times\left(e^{-j\omega\sqrt{r^2+a^2-2ar\cos\psi+z^2}/c_0}-e^{-j\omega z/c_0}\right) \tag{2}$$

in which a is the radius of the circular transducer, $c_0=\sqrt{\frac{K_f}{\rho_0}}$ is the sound speed in air, and where K_f is the incompressibility coefficient. In order to take into account the absorption of ultrasounds in air, it is possible to introduce an effective complex value of the incompressibility coefficient $\tilde{K}_f=K_f(1+j\delta)$, where δ is a dimensionless parameter that represents the damping effect, regardless to the physical mechanism involved.

As the pressure calculus is based on the assumption that every single spot at the surface of the piston radiates a hemi-spherical wave into the medium (i.e. Huygens principle), the field radiated by an annular-shaped source can be deduced by subtracting the field radiated by a disc, whose diameter is equal to the inner diameter of the ring, to the field radiated by a second disc, whose diameter is equal to its outer diameter. Summing up all pressure fields thus calculated for each element of the array, including the central circular piston plus all annular concentric elements, would then provide the acoustic pressure generated by the whole multi-element array, in the frequency domain.

3. Measurement of ultrasonic pressure fields produced by capacitive transducers

In order to characterize the acoustic field generated by an air-coupled transmitter, the experimental technique presented in [9] is used. A 5 μm thick membrane is stretched and placed perpendicularly to the acoustic field at a given distance z away from the source. The incident acoustic wave induces a mechanical movement of the membrane, the normal velocity of which, V_n, can be measured using a Laser Doppler velocimeter. This experimental set up provides the velocity of a point at the surface of the membrane. Then, assuming an incident plane wave and knowing the plane-wave transmission coefficient, T, of the membrane immersed in air, the acoustic pressure P at the insonified surface of the membrane is deduced as follows:

$$P_a = \frac{Z_a}{T} V_n \tag{3}$$

where Z_a=420 Rayl is the acoustic impedance in air. The transmission coefficient T expresses as:

$$T = \frac{4Z_a Z}{(Z_a + Z)^2 \exp(-ikh) - (Z_a - Z)^2 \exp(ikh)} \tag{4}$$

where Z is the acoustic impedance of the membrane, Z=1. 82 MRayl, h is the membrane thickness and k is the wave number of longitudinal wave in the membrane.

In order to check the validity of the model, the first experiment is carried out with a single, circular (radius a=24 mm), capacitive transducer. The input electric signal is a Gaussian-windowed, 10-cycle, 100 kHz centre frequency, sine toneburst. The central frequency is chosen in order to create a Fresnel distance value equal to D_f=168 mm that allows the observation of both near field and far field areas within the measurement range. Besides, the toneburst duration is chosen long enough in order to lead to a quasi-monochromatic regime, but short enough to avoid the temporal superposition of the incident wave and waves reflected between the membrane and the transducer surface.

Figure 2 compares the measured, on-axis, pressure distribution together with the corresponding distribution predicted with and without considering damping in air. The experimental Fresnel distance matches the theoretical one. By considering the experimental amplitude decrease in the far field area, one can observe that for such a frequency the damping effects are not negligible and should be considered in the model for proper prediction. Thus, in the following analysis, these effects are systematically incorporated in the model. The analytical far-field amplitude distribution matches well with the experimental results. Conversely, in the near-field area, discrepancies remain. The reason may lie on the non-uniform movement of the transducer membrane in our experiments. In order to investigate this assumption, the velocity of the surface of the capacitive transducer, i.e. of its membrane, has been scanned using the Laser Doppler velocimeter over a 30 mm long side square area. The measured velocity profile is presented in **Figure 3**. The capacitive transducer surface does not vibrate uniformly because: (1) the backing plate has been sand-blasted so that small cavities of various sizes resonate at its surface with different frequencies and amplitudes, and (2) the membrane does not move as a piston because of its thinness and flexibility and because it is clamped at its edge. Therefore, the velocity, which is measured by the laser probe, is not necessarily the normal-to-the-membrane component, but the projection along the direction of the laser beam of that normal component. This makes a difference with the assumption of piston-like motion of the source, which is introduced in the analytical model, so justifying the disagreement obtained between both measured and predicted pressure amplitude distributions, in the near field area. Moreover, for technical reasons, no measurement was possible between z=0 mm and z=30 mm, because placing the membrane too close to the transmitter was causing reflections between that membrane and the edges of that transmitter, which are protruding with respect to its active surface. Consequently, it was not possible to measure the very near field pressure produced by the transmitter, without strong perturbations caused by overlapping echoes. However, as the far field distribution is based on the total and global contribution of the vibration of the transducer, the experimental measurement is in good agreement with the prediction in far field, which in our case validates the model and provides confidence for its use for further developing NDT applications.

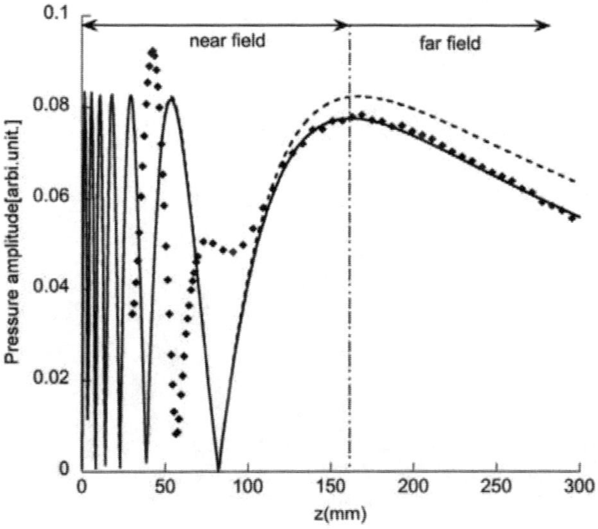

Figure 2. On-axis pressure amplitude radiated by a mono-element, circular transducer of r= 24 mm. Experimental (diamonds) and analytical results with (solid line) and without (dashed line) damping effects.

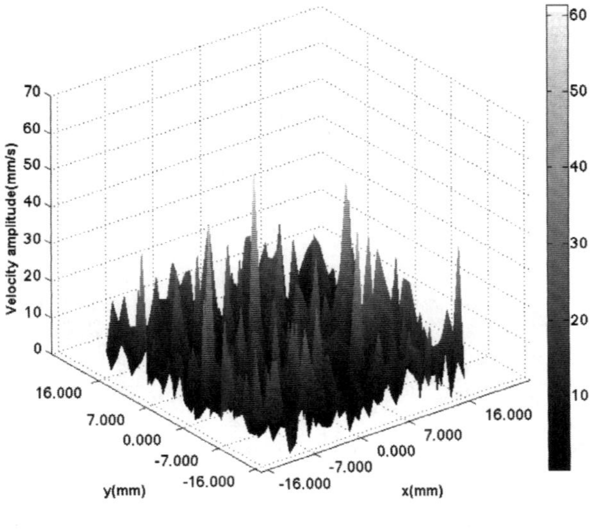

Figure 3. 2D scans of the velocity profile (measured using Laser Doppler velocimeter) of the membrane surface of the mono-element, circular, capacitive transducer (x = $r \cos \psi$ *and* y= $r \sin \psi$).

The case of a wave radiated by one annular element is also considered. This annular element, with inner and outer radius equal to 7 mm and 9 mm respectively, is excited by a Gaussian-windowed, 10-cycle, sine toneburst, while the centre frequency is now equal to 250 kHz. Analytically, the resulting pressure field produced by the annular element is obtained by subtracting the inner field from the outer one. The resulting on-axis pressure amplitude versus the propagation distance z is compared to the experimentally measured one (**Figure 4**).

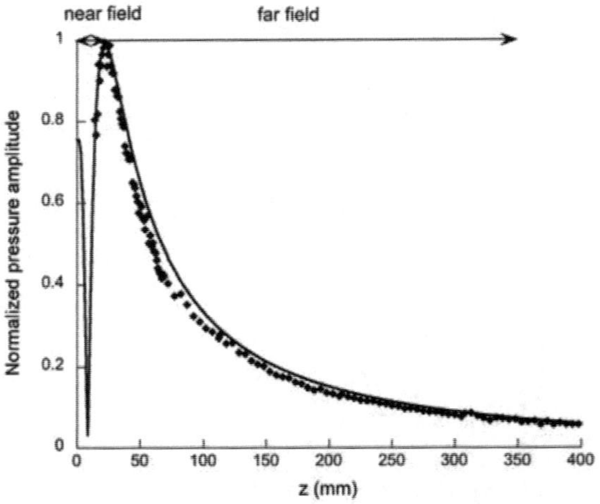

Figure 4. On-axis normalized-amplitude distribution of the pressure field radiated by a single annular element. Experimental (diamonds) and analytical (solid line) results.

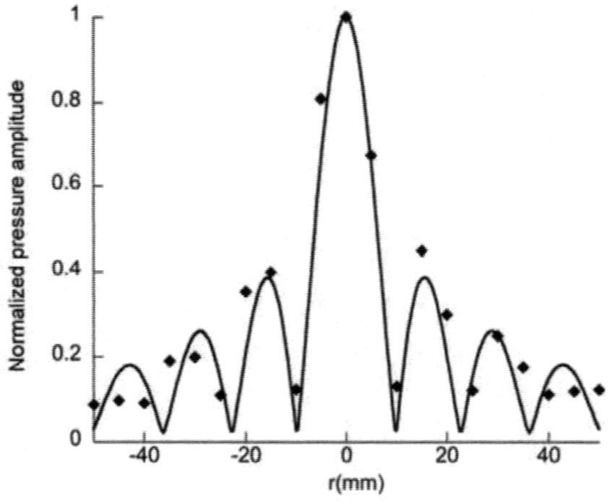

Figure 5. Transverse distribution of pressure amplitude at a distance z=150mm, radiated by the single annular element of inner and outer radius equal to 7 mm and 9 mm, respectively. Experimental (diamonds) and analytical (solid line) results.

Figure 5 compares together both measured and calculated transverse distributions of the pressure amplitude at a distance equal to 150 mm, so in the far field area. The agreement is very good thus validating the modelling approach, which consists of subtracting both acoustic fields predicted for two circular piston sources. This will then be used for modelling the multi-element transducer made up of several annular, concentric elements.

4. Numerical study of multi-element capacitive annular arrays

The case of multi-element arrays is considered analytically, in order to test their focusing properties. Two types of array are studied. Both consist in an arrangement of 8 elements, but with different spatial characteristics. The first one (named "Type I" -**Figure 6**.a) has the following geometrical parameters:

a central circular element with diameter equal to 2 mm is surrounded by 7 annular elements: the spacing between two neighbouring elements is constant and equal to 2 mm, and the width of each element is equal to 2 mm. This leads to a total diameter of the array equal to 58 mm. The second array (named "Type II" -**Figure 6**.b) has the same total outer diameter as the first array, but its elements have all the same area. The central element is circular with a diameter equal to 7.87 mm, and the 7 surrounding annular elements have different widths each other, so that their area keeps the same. For a fixed wavelength, the Fresnel distance depends only on the emitting surface S of the source. Thus, if a continuous wave is applied to each element of this same-area-element array, without any phase delay, the total pressure field will present a natural focal zone.

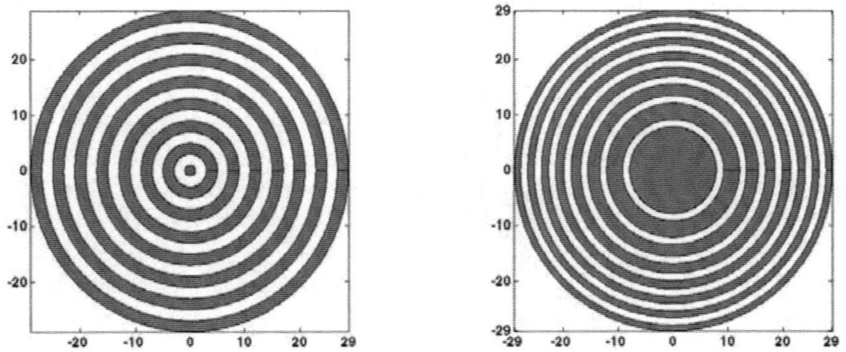

Figure 6. Schematic profile of annular array design a) Type I, where all the elements (in grey) have the same width, and b) Type II, where all the elements share the same area.

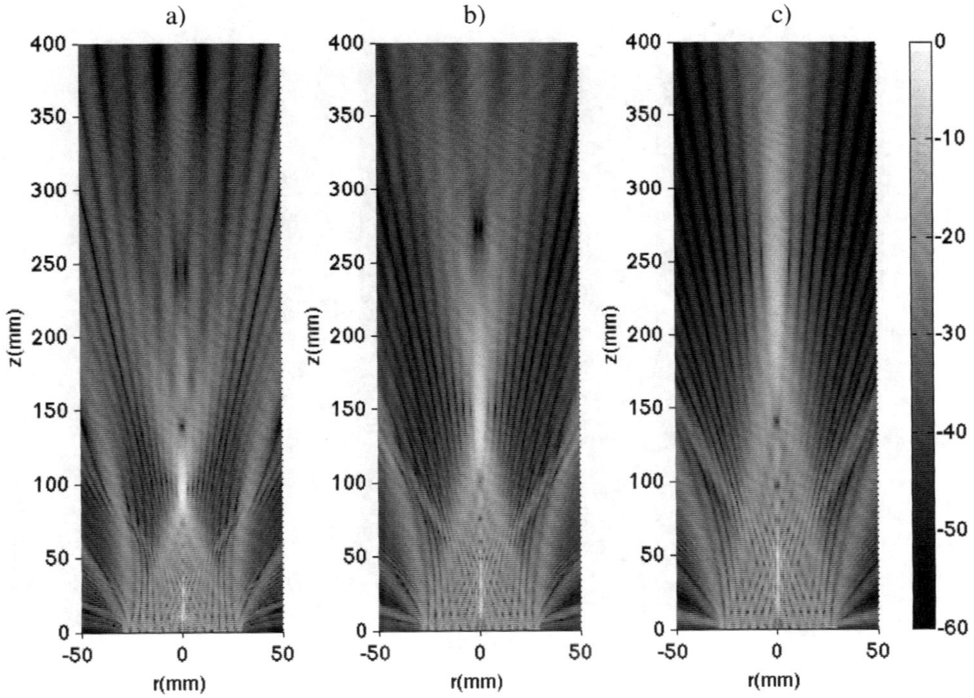

Figure 7. Predicted 2D distribution of acoustic pressure amplitude (in dB scale) produced in air by an 8-element annular array of type I (same width but different areas), focusing at distances: a) z=100mm, b) z=150mm, c) z=200mm.

Journal of Physics: Conference Series **457** (2013) 012011 doi:10.1088/1742-6596/457/1/012011

The pressure fields radiated by these two types of array are calculated. The exciting signal frequency is equal to 250 kHz. The fields radiated by these source (either array of type I or II) are deduced from the fields radiated by their respective central circular element and surrounding annular elements, and phase delays are applied before summing up these various fields together in order to investigate the dynamic focusing capabilities. Predicted distributions, in the (r, z) plane, of the acoustic pressure amplitude radiated by the first array (Type I) are shown in **Figure 7**. Three focal distances are considered: z=100 mm (**Figure** 7.a), z=150 mm (**Figure** 7.b), and z=200 mm (**Figure** 7.c). The pressure amplitude distributions calculated in the same conditions for the case of a Type II array are plotted in **Figure 8** As expected, for both types of multi element arrays, a focal spot appears at the desired location in the air (**Figure 7** and **Figure 8**). The two configurations of annular array design provide similar pressure amplitude at each focal distance, even if the array of type II provides slightly higher amplitude than that produced by the array of type I, at the focusing points. As the field is focused at a large distance, the pressure amplitude decreases and the lateral dimension of the focal spot increases, deteriorating the transverse spatial resolution for further use of the transducer as a materials scanner. Acoustic pressure amplitude distributions in transverse direction are plotted in **Figure 9**, for the three previous focusing distances: z=100 mm, z=150 mm and z=200 mm. For the nearest focal distance (z=100 mm), grating lobes are visible for the type I array, at r = ± 35 mm (**Figure 9**.a). The width of these grating lobes is about 10 mm. Although their amplitude is about 22 dB less than that of the focal spot amplitude, such grating lobes may induce artefacts if involved in NDT measurements. Conversely, for the array of type II, these grating lobes have significantly reduced amplitudes (**Figure 9**.b), thus making this type of array more suitable for NDT applications.

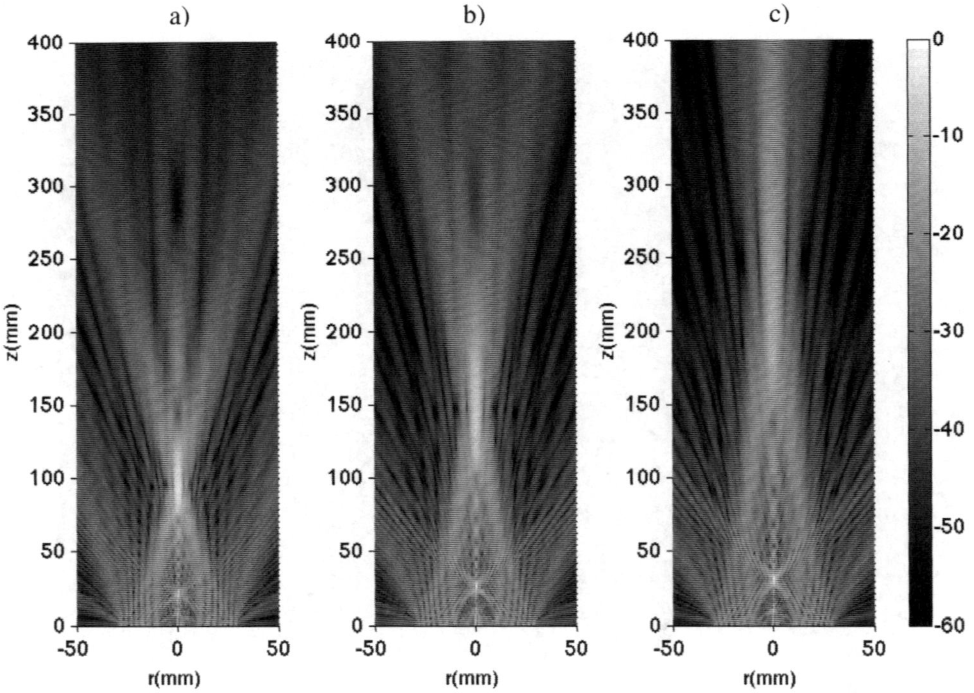

Figure 8. Predicted 2D distribution of acoustic pressure amplitude (in dB scale) produced in air by an 8-element annular array of type II (the same area but different widths) focusing at distances: a) z=100 mm, b) z=150 mm, c) z=200 mm.

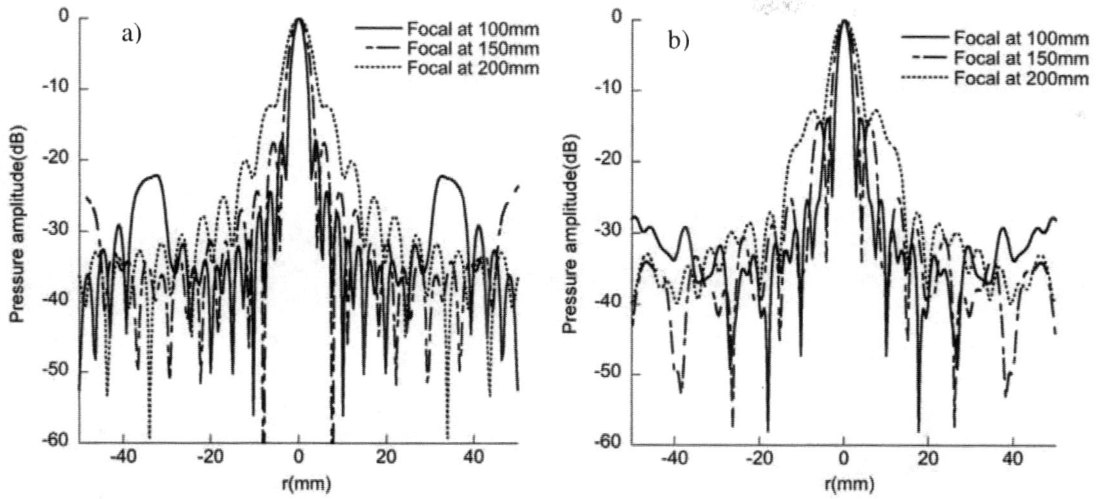

Figure 9. Predicted transverse distribution of pressure amplitude (in dB scale) at a distance z equal to the aimed focussing distance, for both types of 8-element annular arrays: a) type I (the same width but different areas) and b) type II (the same area but different widths).

5. Summary

This paper introduced a multi-element capacitive annular array designed for non-destructive testing (NDT) purposes. An analytical model was built to simulate the acoustic field generated in air by circular or annular, individual elements or by an array made of one central circular element and several surrounding annular elements. The model has been validated through the comparison of several predicted distributions of pressure with measurements carried out using a Laser Doppler velocimeter. Then, using the model, two types of 8-element annular arrays have been numerically investigated and their performances compared together. The first one, was made of one central circular element plus seven annular elements all having the same width (and so different areas); the second one was also made up of one central and seven annular elements but these were designed so that they all have the same area. Dynamic focusing has been numerically shown possible for both arrays, but the array made of elements having all the same area was shown to have no (or very little) grating lobes. This geometrical configuration is therefore more suitable for further NDT applications. Future work will focus on the fabrication of the transducer arrays, and the optimized choice of the filling material between the elements, in order to minimize cross-talking effects.

Acknowledgments
This work was partly funded by the Chinese Scholarship Council.

References

[1] Grandia W A and Fortunko C M 1995 vol 1 *Ultrasonics Symp., 1995. Proc., 1995 IEEE* vol 1 pp 697 –709

[2] Kays R, Demenko A, Maeika L, literis R and ukauskas E 2007 *Insight - Non-Destructive Testing and Condition Monitoring* **49** 195–9

[3] Monchalin J-P 2004 ed D O Thompson, D E Chimenti, L Poore, C Nessa and S Kallsen *AIP Conf. Proc.* **700** 3–31

[4] Gao H, Ali S M and Lopez B 2010 ed D O Thompson and D E Chimenti *AIP Conf. Proc.* **1211** 1175–81

[5] Farlow R and Hayward G *Insight* **36** 926–35
[6] Yañez Y, Garcia-Rodriguez M, Garcia-Hernandez M J, Salazar J, Turo A and Chavez J A 2008 *NDT & E Int.* **41** 292–9
[7] Fortunko C M and Chimenti D E 1993 vol 2 *Ultrasonics Symp., 1993. Proc., IEEE 1993* pp 655–8
[8] Schindel D W and Hutchins D A 1991 *Ultrasonics Symp., 1991. Proc., IEEE 1991* Ultrasonics Symp., 1991. Proc., IEEE 1991 (vol 1 IEEE) pp 301–4
[9] Biateau C, Hosten B and Roziere D 2002 *AIP Conf. Proc.* **615** 921–8
[10] Ergun A S, Yaralioglu G G and Khuri-Yakub B T 2003 *J. Aeros. Eng.* **16** 76–84
[11] Castaings M and Hosten B 2008 *NDT & E Int.* **41** 648 - 55
[12] Anderson M J 1995 *J. Acous. Soc Am,* **97** 262–72
[13] Hietanen J, Mattila P, Stor-Pellinen J, Tsuzuki F, Vaataja H, Sasaki K and Luukkala M 1993 *Meas. Sci. Technol.* **4** 1138–42
[14] Carr H, Munro W S H, Rafiq M, and Wykes C, 1992 *NDT& E Int,* **10** 3–13
[15] Mattila P, Tsuzuki F, Vaataja H and Sasaki K 1995 *IEEE Trans. Ultrason. Ferroelect. Freq. Contr.* **42** 1–7.
[16] Hietanen J, Stor-Pellinen J and Luukkala M 1992 *Meas. Sci. Technol.* **3** 1095–7
[17] Pizarro L, Certon D, Lethiecq M and Hosten B 1999 *Ultrasonics* **37** 493–503
[18] Schindel D W, Bashford A G and Hutchins D A 1997 *Ultrasonics* **35** 275–85
[19] Song J and Chimenti D E 2006 *Int. J. Appl. Sci. Eng.* **4**.1 1-19
[20] Octavio Manzanares A and Montero de Espinosa F 2012 *Ultrasonics* **52** 482–9
[21] McGough R J, Samulski T V and Kelly J F 2004 *J. Acoust. Soc. Am.* **115,** 1942-54

AUTHOR INDEX

Institute of Physics
Dirac House, Temple Back
Bristol BS1 6BE UK

CURRAN ASSOCIATES INC
proceedings.com